Internal Corrosion Control of Water Supply Systems
Code of Practice

Metals and Related Substances in Drinking Water
Code of Practice Series

Internal Corrosion Control of Water Supply Systems
Code of Practice

Edited by

Dr. Colin Hayes

Publishing

London • New York

Published by   IWA Publishing
                 Alliance House
                 12 Caxton Street
                 London SW1H 0QS, UK
                 Telephone: +44 (0)20 7654 5500
                 Fax: +44 (0)20 7654 5555
                 Email: publications@iwap.co.uk
                 Web: www.iwapublishing.com

First published 2012
© 2012 IWA Publishing

British Library Cataloguing in Publication Data
A CIP catalogue record for this book is available from the British Library

ISBN 9781780404547 (Paperback)
ISBN 9781780404554 (eBook)

Contents

About the Code of Practice

This Code of Practice is part of a series of publications by the IWA Specialist Group on Metals and Related Substances in Drinking Water. It complements the following IWA Specialist Group publications:

- Best Practice Guide on the Control of Lead in Drinking Water
- Best Practice Guide on the Sampling and Monitoring of Metals in Drinking Water
- Guide for Small Community Water Suppliers and Local Health Officials on Lead in Drinking Water

The Code of Practice is concerned with metal pick-up by drinking water within the water supply chain, particularly copper, iron and lead, and to a lesser extent nickel and zinc. The emphasis is on cold drinking water at its point of use by consumers.

The intention is that this Code of Practice establishes an international standard for the control of internal corrosion of water supply systems. It has been produced by an international panel of authors with representation from regulatory agencies, water utilities, professional institutions, technical consultancies, research institutes and universities. It has also been subjected to wide-spread peer review.

The Code of Practice will be reviewed and the e-book version updated on an annual basis to ensure that it keeps up to date with the needs of the user. It will be particularly important to adapt to changes in regulations, some of which are likely to become more stringent.

Authors

Dr. Angelika Becker, IWW (DE)
Maria Benoliel, EPAL (PT)
Matthew Bower, Drinking Water Quality Regulator for Scotland (UK)
Andy Campbell, City of Ottawa Water (CA)
Dr. Brian Croll, WQM Associates Ltd (UK)
Paul Gadoury, Providence Water (US)
John Griggs, Chartered Institute of Plumbing and Heating Engineering (UK)
Prof. Xiaohong Guan, Tongji University (CN)
Dr. Colin Hayes, WQM Associates Ltd and Swansea University (UK)
Dr. Dragana Jovanovic, Institute of Public Health of Serbia (RS)
Martin Jung, Austrian Institute of Technology (AT)
Dr. Yi-Pin Lin, National University of Singapore (SG)
Dr. Darren Lytle, US Environmental Protection Agency (US)
Prof. Michael Ritchie Moore, Water Quality Research Australia (AU)
Dr. Adam Postawa, AGH University of Science and Technology (PL)
Prof. Michele Prevost, Montreal Polytechnic (CA)
Dr. Larry Russell, Reed International (US)
Michael Schock, US Environmental Protection Agency (US)
Dr. Daniel Tsang, Hong Kong Polytechnic University (HK)
Dr. Simoni Triantafyllidou, Virginia Tech (US)

Review Panel

The authors and publisher wish to thank all those involved in the review of this Code of Practice. The review involved the Management Committee of the IWA Specialist Group on Metals and Related Substances in Drinking Water and the following:

Clayton Commons, Rhode Island Health Department (US)
Copper Development Association (UK)
Dr. Eddo Hoekstra, European Commission, Joint Research Centre (INT)
Owen Hydes OBE, formerly Drinking Water Inspectorate (UK)
National Health and Medical Research Council, Lead Working Committee (AU)
David Scott, Toronto Water (CA)
Technical Committee of Chartered Institute of Plumbing and Heating Engineering (UK)

Glossary

Cuprosolvency: The tendency of water to dissolve copper.

DIC: Dissolved organic carbon.

Galvanic corrosion: Electrochemically induced corrosion at the site of two different metals.

Legacy leaded solder: Solder containing appreciable amounts of lead (often $> 8\%$) that was used prior to the mid 1980s.

Plumbosolvency: The tendency of water to dissolve lead.

Sequential sampling: A series of samples taken in sequence, typically after water has stagnated within pipework.

RDT sampling: Random daytime sampling.

Risk assessment: As applied to internal corrosion control, the systematic assessment of risks from corrosion throughout the entire water supply systems.

Tuberculation: Voluminous corrosion deposits that can often block pipes.

XRD: X-ray diffraction.

Disclaimers

Whilst every reasonable attempt has been made to present the information in this Code of Practice in a fair and balanced manner, the reader should none-the-less satisfy themselves of its relevance to their specific circumstances. It must also be appreciated that some aspects of the topic of internal corrosion control do not enjoy total consensus of opinion and that practices have varied around the world.

The mention of specific companies or of certain manufacturers' products does not imply that they are endorsed or recommended by the publisher or authors in preference to others of a similar nature that are not mentioned. All reasonable precautions have been taken by the authors to verify the information contained in this publication. However, the published material is being distributed without warranty of any kind, either express or implied. The responsibility for the interpretation and use of the material lies with the reader. In no event shall the publisher or authors be liable for damages arising from its use. The views expressed by the authors do not necessarily represent the decisions or the stated policies of any organization referred to in this publication.

The U.S. Environmental Protection Agency, through its Office of Research and Development, funded and managed, or partially funded and collaborated in, the research described herein. It has been subjected to the Agency's administrative review and has been approved for external publication. Any opinions expressed in this Code of Practice are those of the authors and do not necessarily reflect the views of the Agency, therefore, no official endorsement should be inferred. Any mention of trade names or commercial products does not constitute endorsement or recommendation for use.

Foreword

A Code of Practice for the management of corrosion in water distribution systems is important both economically and for protection of public health. When I started to study plumbosolvency 40 years ago (Moore, 1973), I did not realise where it would end up. I was not the first to perceive this issue. A century previously, Robert Christison in Edinburgh sounded the warning bells about the proposed new water supply to Glasgow (Christison, 1844). His fears were assuaged by Robert Stephenson and Isambard Brunel who believed that exposure, "would not be injurious". However, further in the past, others like James Lind recognised that acidic solutions would dissolve lead from glaze on earthenware vessels (Lind, 1754). In retrospect it should have been obvious that a soft, acidic, unbuffered water supply would dissolve lead from the distribution piping. Analytical technology was then, a century and more ago, inadequate for the task. Gravimetric methods could not deal with the very small quantities involved. As we now luxuriate in the Pico molar technology of ICPMS and the like, we should remember that just 50 years ago we were still measuring lead (and other metals) by dithizone extraction and polarography. The sub-clinical health issues associated with long-term over-exposure to lead have only unfolded over the past 40 years. The lessons of that time and subsequently are that drinking water can be a primary vector of exposure to a number of solutes, including lead, with potential health consequences.

The insidious nature of this type of exposure from drinking water is still likely to be ignored in the face of more strident cries for attention from other exposures (World Health Organization, 2010). Low-level exposure to lead of this nature does not cause acute, fulminant illness like water-borne pathogens. Rather, it is associated with slow chronic onset of sub-clinical disease that impacts on populations rather than individuals. These exposures to low-level contaminants creep like thieves in the night into our homes causing subtle, barely recognisable, decrements in ability and function. This Code of Practice provides the world water community with guidance on and a framework for corrosion control, predicated on the best international knowledge, skills and experience – an international standard. Adoption of these protocols will allow water supply authorities to mitigate the economic and public health risks associated with the distribution of potable water supplies. The Code minimises possible exposure to a range of metallic elements including those, like lead, associated with impoverishment of human health. As an exposure vector our potable water supplies hold a unique position in their capacity to have major population health impacts where water quality is inadequately controlled. The unanticipated results of exposures to metals can be managed by following the recommendations of the Code of Practice thus avoiding the unintended public health consequences of corrosion.

Christison, R. (1844). On the action of water on lead. Trans. Roy. Soc. Edin. 15 265–276.

Lind, J. (1754). On the danger of using certain earthen vessels. Scots Mag. 16, 227–229.

Moore, M. R. (1973). Plumbosolvency of waters. Nature 243, 222–223.

World Health Organization. (2010). Booklet on Childhood Lead Poisoning. ISBN 978 924 150033 3.

Michael R Moore

Part A
Code of Practice

Chapter 1

Introduction

1.1 SCOPE

This Code of Practice is concerned with metal pick-up by drinking water within the water supply chain, particularly from water mains and from domestic and institutional pipe-work systems. The principal metals of interest are copper, iron, and lead, and to a lesser extent nickel and zinc. The emphasis is on cold drinking water at its point of use by consumers. Metals arising from water sources and hot water systems are not considered.

1.2 PURPOSE

The intention is that this Code of Practice establishes an international standard for the control of internal corrosion of water supply systems. It provides a basis for identifying both problems and sustainable solutions in a manner which is sound scientifically and will help operators to achieve due diligence. It provides a template for improving internal corrosion control in countries, cities or towns where this has been neglected or poorly implemented.

1.3 APPLICATION AND CONTENT

The Code of Practice is deliberately brief in its presentation of a wide array of complex information, in order to provide direction to practitioners that can be more easily related to their specific circumstances. Its focus is on the application of key principles in five stages, as illustrated in Figure 1.1.

Part B, Supporting Information, provides brief guidance on technical issues. Key references are provided throughout to signpost further more detailed information.

Part C provides a series of check-lists and criteria to be used in risk assessment.

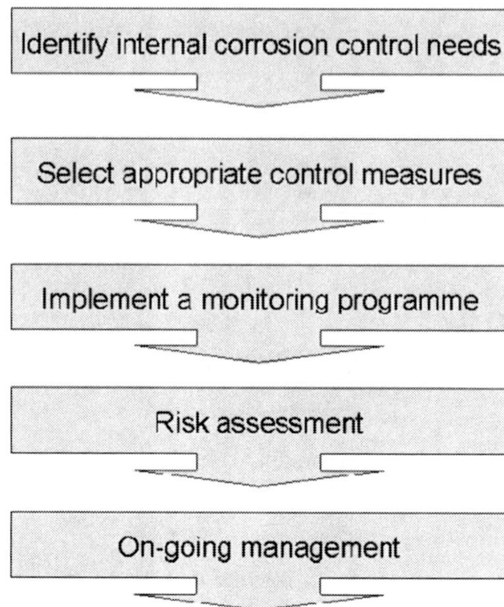

Figure 1.1 The five stages in the Code of Practice.

Chapter 2

Identifying internal corrosion control needs

2.1 GAINING A SYSTEM OVERVIEW

A diagnostic assessment, from "source to tap" should first be undertaken to ensure that overall perspectives are appreciated, as outlined in Table 2.1. The diagnostic assessment should aim to identify the possible extent and location of any internal corrosion problems and their potential causes.

Table 2.1 Diagnostic assessment.

Points to consider	Relevance
Size and complexity of the water supply system	Solutions may need to be system-wide or could be localised; multiple points of treatment may be necessary
Raw water source(s): alkalinity (DIC), pH, organic quality, particularly humic and fulvic acids (colour)	Potential for corrosion increases with lower pH (particularly with lower alkalinity) and higher organic contents
Extent and reliability of water treatment available	Is treatment effective in reducing corrosion potential?
Disinfection regime and whether Pb (II) or Pb(IV) compounds are governing lead solubility	The nature and concentration of disinfectant residuals can influence internal corrosion and metal release
Nitrification in chloraminated systems	May lower pH in poorly buffered waters
Water quality throughout the distribution network	Changes in water quality may affect corrosion, such as pH or the ratio of chloride to sulphate

(Continued)

Table 2.1 Diagnostic assessment (*Continued*).

Points to consider	Relevance
Water supply temperature variation	Generally, corrosion is greater at higher temperatures
Extent of cast-iron water mains in use and their condition	Will determine the potential need for replacement or rehabilitation
Number of lead service pipes in service and extent of any lead pipes in homes/buildings	Will largely determine the need for plumbosolvency control
Extent of use of brass fittings and brass pipes	May influence the choice of pH regime to minimise dezincification. May also be relevant to plumbosolvency control.
Historic and any current use of solders containing lead	Potentially relevant to minimising lead and could restrict coagulant changes in water treatment.
Number of copper service pipes in service and extent of any copper pipes in homes/buildings.	Will largely determine the need for cuprosolvency control
Number of galvanised iron service pipes in service and extent of any iron pipes in homes/buildings	May influence corrosion control
Extent of use of chrome-nickel taps and fitting	May influence corrosion control
Use of corrosion inhibitors: type and concentrations used	May need to be optimised
Current programmes for replacing or rehabilitating water mains and service pipes	May need to be optimised
Control of the metal fittings used in the system	May need to be optimised and can be more economical than treatment in small water systems
Complaints from consumers	May indicate either localised or system-wide problems: see Section 2.2
Regulatory compliance	Normally the principal driver for internal corrosion control: see Section 2.3

2.2 EVIDENCE OF INTERNAL CORROSION PROBLEMS

Internal corrosion problems are likely in the circumstances summarised in Table 2.2. The circumstances that can cause or exacerbate internal corrosion problems should be investigated periodically and preferably as part of an annual risk assessment (see Section 5).

Table 2.2 Circumstances associated with corrosion problems.

Circumstances	Likelihood of corrosion problems
Water mains at the extremity of the distribution network	Iron release from old cast-iron water mains will be exacerbated when water stagnates.
Microbiological growth within the distribution network is evident from elevated colony counts or coliforms	Iron release from old cast-iron water mains will be exacerbated by active bio-films within mains and their corrosion tubercles. Nitrification induced pH reductions in poorly buffered waters can increase copper and lead releases.
Long pipe runs, particularly in institutional buildings	Elevated copper release, especially from new copper pipes, will increase under prolonged water stagnation when oxidants persist.
Non-residential buildings with intermittent water use	Risks of elevated lead release may be higher from all potential sources.
Degradation of brass fittings and legacy leaded solder	Particulate lead may build up in tap filters.
pH in water supplies is <7 and particularly <6.5	Acidic water is corrosive to all metals in contact, particularly cast iron, galvanised iron, copper and lead (the more so at lower pH values).
Chloride/alkalinity ratio is >0.5 at pH >8.3	High risk of meringue dezincification of brass fittings.
Chloride/sulphate ratio is >0.6	Galvanic corrosion of legacy leaded solder is more likely.
Water treated by reverse osmosis is not conditioned to increase alkalinity	Treated waters with low ionic contents can be corrosive to all metals in contact if without proper post-treatment.
Corrosion inhibitors are not used	pH adjustment alone may be insufficient.
Silicates are dosed for corrosion control	Only relevant to iron discolouration control.
Polyphosphates are dosed for corrosion control	Whilst polyphosphates can be beneficial for mitigating iron discolouration, other metal releases (Cu, Pb) may increase.
Orthophosphate is dosed for plumbosolvency control	Corrosion inhibition is dose, alkalinity and pH dependent. Low doses are likely to be ineffective.
Changes in water sources from groundwater to surface water	Iron release from the corrosion tubercles in old cast-iron water mains.
Elevated metal concentrations have been detected in blood samples	Possible evidence of corrosion problems, albeit sources other than drinking water may be responsible.

Direct evidence of internal corrosion problems may arise from consumer complaints, routine system maintenance, regulatory compliance assessment or investigations. Table 2.3 summarises the most common types of evidence that may be found. Problems with lead and nickel are normally not evident visually or detected aesthetically by consumers.

Table 2.3 Direct evidence of internal corrosion problems.

Evidence	Nature of Problem
Consumer complaints of discolouration (red water)	Occasional complaints suggest only a local problem with iron. Regular complaints suggest that corrosion of cast-iron water mains is significant
Sections of cast-iron water mains, cut out from a repair, show internal tuberculation (see Figure 2.1)	Clear evidence of iron corrosion but needs to be put into perspective (extent and location)
Filter based samplers from water mains collect significant amounts of iron-rust particles	Clear evidence of corrosion in the vicinity of the water mains being investigated
Consumer complaints of green or blue staining of washing or sanitary-ware	Indicative of general copper pick-up which will be more evident with longer copper pipes and prolonged stagnation times
Seepage from copper pipework (see Figure 2.2)	Indicative of copper pitting corrosion
Sections of copper pipe (cut-outs) show irregular mounds of corrosion deposit (see Figure 2.3)	Indicative of copper pitting corrosion
Consumer complaints of "sandy" water	Indicative of dezincification of brass fittings
Physical failure of brass fittings	Indicative of dezincification of brass fittings
Particles have been trapped in faucet aerator screens	May be indicative of particulate lead
Non-compliance with standards for metals in drinking water	Whilst this is the most common evidence available, caution is required in the interpretation of results (see Section 2.3 and Part B, Appendix 1)

Manganese can cause "black water" discolouration problems, arising normally from the natural contamination of source water and its precipitation and subsequent mobilisation (by physical disturbance) within a distribution network; there are also emerging health concerns. Manganese control will involve source water treatment (typically oxidation and filtration) and flushing of the distribution network; it is not directly relevant to corrosion control except that manganese can co-precipitate with iron corrosion products, which may absorb lead when in contact with lead pipework.

Figure 2.1 Section through a heavily corroded cast iron water main (courtesy of Dr Darren Lytle, US EPA).

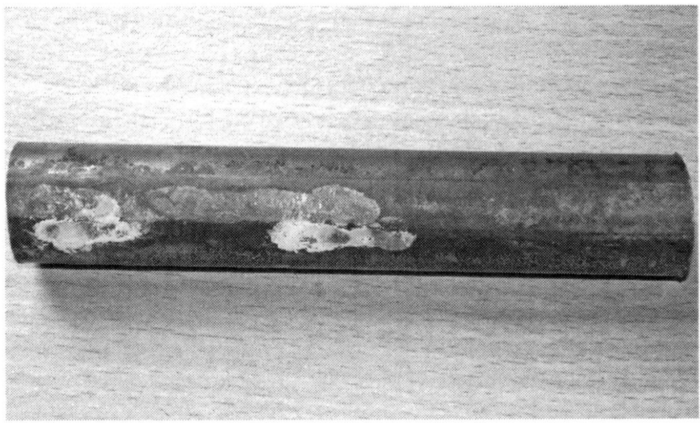

Figure 2.2 Copper pipe showing signs of leakage due to localised corrosion (courtesy of Steve Hepple, Swansea University).

Figure 2.3 Internal surface of a corroded copper pipe (courtesy of John Griggs, CIPHE).

Indirect evidence of internal corrosion can be gained from coupon or pipe-rig studies, as outlined in Part B, Appendix 2. An immediate dilemma is whether to use new or old exhumed pipes for testing purposes. New lead pipes have been shown to provide a reliable basis for characterising plumbosolvency (when Pb(II) compounds are involved) and test results can be obtained within a month. Old in-situ lead pipes can take up to several years to equilibrate with a new water quality condition (e.g. a step change in orthophosphate dose) making optimisation potentially difficult when using old lead pipes in test rigs (IWA, 2010). In water supply systems that carry high free chlorine residuals, Pb(II) corrosion deposits may convert over time to Pb(IV) corrosion deposits that have a significantly lower solubility, negating short-term testing techniques.

Corrosion testing for copper is also complicated by changes to the corrosion deposits that occur over time. Initially, cuprous or cupric oxides form, but there is normally a gradual conversion to copper carbonate that has a lower solubility. In consequence, cuprosolvency effects differ with the age of copper pipework. For the characterisation of general copper corrosion, test rigs can be deployed within a water supply system, typically using a set of pipes with different water stagnation times, with routine sampling as a normal operational function. Test rigs can be automated to improve sampling efficiency and may include brass fittings and other pipe materials if broader metal release is of interest. A key consideration should be the representativeness of any pipe rigs used in relation to service conditions, which can change, such as when on-site treatment is installed (e.g. water softening).

Particularly when water quality conditions are more extreme, such as with low pH or high salinity, shorter term pipe tests should be able to determine the propensity for copper pitting corrosion. In such cases, sections of copper pipe will require sacrificial examination over the time span of testing, likely to be several months. Corrosion deposits from testing or from old exhumed pipes can be examined by a range of techniques to determine their physical and chemical nature (see Part B, Appendix 2 and AWWA, 2011).

2.3 REGULATORY COMPLIANCE

Regulatory standards for metals in drinking water are either based on aesthetic or health considerations and differ to some extent across the world, both in terms of numeric values and the methods of compliance assessment. The sampling methods used are not without problems and in consequence there is a danger that compliance results can be misleading. Such issues, including the representativeness of the sampling methods being used, must be considered in routine risk assessment (Section 5). In Part B, Appendix 1 summarises sampling characteristics and how these affect compliance, and Appendix 3 outlines the compliance modelling techniques that can help Utilities to investigate the relationship between corrosion control treatment conditions and likely regulatory compliance. Regulatory compliance tends to focus on the position across an entire water supply system. Assessing copper, lead or nickel emissions at an individual property will be best achieved by sequential stagnation sampling (30 minutes stagnation samples will be adequate for lead) or by split-flow composite sampling (IWA, 2012). Samples taken during flushing can be used to investigate particulate metal problems.

Chapter 3
Selecting appropriate control measures

3.1 PLANNING

Internal corrosion control for a water supply system should be planned. All aspects should be considered and investigated if necessary, including pH and its stability, the potential for effects from organics and particulate iron, and the suitability and extent of optimisation of any corrosion inhibitors in use. Internal corrosion control needs to be integrated into the development of overall water quality objectives, because the control of disinfection and other regulated contaminants may necessitate other concurrent treatment processes to assure the highest quality water to the consumer. Plans should be reviewed periodically and kept up to date.

It is important to appreciate that there is no universal solution to internal corrosion problems. For example, some measures to control iron corrosion in old cast-iron water mains are different (and can even be contradictory) to those needed to reduce lead. As part of the overall planning and commitment to an optimised corrosion control programme, the operators of a well-run water supply system should have a good understanding of the corrosion control mechanisms that relate to their water quality and pipe materials. This knowledge will help to avoid unintended consequences from operational decisions and also to define appropriate optimisation tests and their conditions.

Economic analysis is essential in order to determine the optimum range of measures that may be warranted in a water supply system. For example, it may be economic to replace all lead pipes in a system (including those owned by house-holders) if numbers are small, compared to installing, maintaining and operating centralised treatment facilities. Such an approach may encounter ownership and liability issues, however. Where the number of lead pipes is high, centralised treatment is likely to be far less expensive than wide-spread replacement (IWA, 2010), although the total removal of lead pipes must be the ultimate goal.

Expenditure may need to be phased, depending on the extent of work required and the severity of problems. Particularly with iron corrosion, investigations may need to be substantial if optimum measures are to be identified. The costs associated with investigations should be fully incorporated into expenditure planning.

With very small water supplies, particularly when privately owned, treatment measures are best avoided, as the necessary expertise is unlikely to be available. Even so, certified small-scale water treatment units from reputable suppliers can be considered, particularly if maintained by appropriately trained and knowledgeable personnel.

3.2 pH OPTIMISATION

It is fundamentally important to optimise pH within a water supply system. On no account should acidic water with a pH below 6.5 be passed through a distribution network or domestic or institutional pipe-work system that contains metal materials, and preferably, pH should be kept above 7.0.

Typical and generally acceptable pH ranges for the main water types are as follows:

- Ground-waters: pH 7.0 to 7.5
- High alkalinity surface waters: pH 7.5 to 8.0
- Low alkalinity surface waters: pH 8.0 to 9.0

These pH ranges reflect traditional good practice and a broad approach to corrosion control in which iron discolouration control is an important consideration, particularly in Europe. In the US and Canada, it has been long-established practice in some water supply systems to elevate the pH of low alkalinity waters up to 10.0.

The pH range 6.5 to 8.5 is a regulatory requirement in many countries (although 6.5 to 9.5 in Europe), with variable levels of enforcement, or it forms the basis of operational guidelines. In the US and EU, pH standards are qualified by a requirement that drinking water should be "non corrosive" or "not aggressive", respectively (see Part B, Appendix 1). The most common emphasis has been the pH elevation of low alkalinity waters which have the highest potential for corrosivity.

For specific corrosion control purposes, a higher pH can be beneficial, particularly for low alkalinity waters where a pH of 9.0 to 10.0 can be used to suppress plumbosolvency. Consideration needs to be given, however, to regulatory standards for pH – for example in the UK, the upper limit at consumer's taps is 9.5. Water may comply with this standard as it leaves the treatment works, but the presence of cementitious materials in the distribution system may further increase pH and cause compliance issues. The presence of nitrifying biofilms may lower pH during distribution, complicating pH control.

Increasing the pH of high alkalinity waters to 8.5 (or above) for plumbosolvency control may require centralised water softening, to avoid calcite precipitation problems, as widely practised in the Netherlands. However, many water utilities have preferred to dose orthophosphate to high alkalinity waters rather than attempt to increase the pH beyond 8.0.

The optimum pH for plumbosolvency control will vary when orthophosphate corrosion inhibitor is dosed, depending on the extent of natural organic matter (NOM) present; generally, the pH needs to be higher as the amount of NOM that is present increases. The pH conditions used in the control of plumbosolvency should also be applicable generally to the control of copper and iron corrosion.

Dezincification of brass fittings can be a problem if the chloride (mg/l) to alkalinity (mg/l as $CaCO_3$) ratio exceeds 0.5, particularly with high-zinc "yellow" brasses. High-copper "red" brasses are more resistant to dezincification but their higher cost has minimised their use. If the pH is higher than 8.3 then "meringue" dezincification can occur where the white zinc corrosion products precipitate at the site of corrosion and may consequently block fittings.

High alkalinity waters have a good buffering capacity and pH changes throughout a water supply system should be minimal. Low alkalinity waters less than 50 mg/l (as $CaCO_3$), and particularly less than 20 mg/l (as $CaCO_3$), will be prone to pH reductions as a consequence of nitrification or carbonic acid generation by bio-films within pipes. In such cases, an increase in buffering capacity should be considered, for example by dosing lime in conjunction with carbon dioxide, or by filtration through dolomitic limestone, to increase alkalinity.

3.3 SELECTING A CORROSION INHIBITOR

To reduce the effects of iron corrosion, either silicates or polyphosphates can be dosed. These chemicals work by binding with ferrous ions at the sites of corrosion, thereby avoiding oxidation and precipitation, and keeping these ions in solution. They have limited effect on pre-formed particulate ferric compounds. Caution is needed when using polyphosphates as they can also dissolve other metals, notably lead.

The most effective corrosion inhibitor for suppressing plumbosolvency is orthophosphate. Suppression is dose dependent and it is therefore essential to establish the dose required that will fulfil the required reductions in lead (IWA, 2010). To suppress cuprosolvency, orthophosphate should also be effective at the doses used in plumbosolvency control; this appears to be the case in England and Wales where 95% of supplies are dosed with orthophosphate (optimised for plumbosolvency control) and >99.9% of random samples were compliant with the copper standard of 2 mg/l during 2009 and 2010.

There is circumstantial evidence from the UK, from the compliance data, that orthophosphate suppresses nickel solvency from chrome-nickel plated fittings and suppresses lead leaching from brass and legacy leaded solder. In some countries, utilities prefer to use zinc orthophosphate, rather than orthophosphoric acid. Laboratory testing, the use of pipe-rigs and pilot trials should all be considered to ensure the correct selection of corrosion inhibitor.

3.4 OPTIMISING THE DOSING OF CORROSION INHIBITORS

The optimisation of orthophosphate dosing for plumbosolvency control can be achieved in one of several ways, for example:

(1) Step changes in orthophosphate dose, based on professional judgement, until optimum reductions in lead have been demonstrated by lead pipe-rigs or in-situ lead pipes at consumers' houses, both likely to involve a stagnation sampling method; however, it is important to appreciate that old lead pipes can take up to two or three years to equilibrate to a new orthophosphate dose (IWA, 2010) and in consequence the potential benefit of a particular treatment condition might not be fully demonstrated;

(2) Laboratory testing (Part B, Appendix 2) coupled with compliance modelling (Part B, Appendix 3) to quickly determine the optimum dose, which must then be confirmed (and adjusted if necessary) by routine monitoring of in-situ lead pipes at consumers' houses; this approach can minimise the number of iterative changes to water treatment conditions and can save both time and expenditure; or

(3) The application of a generic dose based on water type; this simplistic approach may result in under-dosing that does not satisfy compliance objectives or over-dosing at higher cost, but it may be the only realistic way of dealing with very small water supply systems.

As a guide, the optimum dose ranges of orthophosphate for plumbosolvency control with the main water types will be as follows, on the basis of widespread UK practice which aims to reduce lead below 10 µg/l:

- Ground-waters: 1.0 to 1.5 mg/l (P);
- High alkalinity surface waters: 1.0 to 1.5 mg/l (P), but up to 2.0 mg/l (P) if organics are significant;
- Low alkalinity surface waters: 0.5 to 1.0 mg/l (P), but up to 2.0 mg/l (P) if organics are significant.

Orthophosphate doses may also need to be higher if iron, manganese, aluminium or calcium precipitates are encountered within the distribution network.

The above dose ranges are applicable to lead release from lead service lines and internal lead plumbing and, on the basis of circumstantial evidence from the UK, also appear to be applicable to suppressing lead releases from brass and from secondary pipe deposits involving lead with iron, aluminium or manganese. It is important that any interference effects from organics and/or iron particulates are dealt with as part of a holistic corrosion control strategy.

It can be noted that 1 mg/l orthophosphate as P is equivalent to 3 mg/l as PO_4.

Different lead standards will be associated with their own appropriate optimum orthophosphate dose and optimum pH, and it should also be noted that orthophosphate doses are system specific, dependent on the water quality and the circumstances of the water supply system. By implication, optimum orthophosphate doses and optimum pH should be determined specifically for individual water supply systems. In these respects, a generic definition of the term "optimisation" is provided in Part B, Appendix 4 with three protocols for achieving optimisation of plumbosolvency control in Part B, Appendix 5.

For larger water supply systems serving more than 100,000 people, there are no compelling arguments for not determining optimum plumbosolvency control conditions on a specific and scientifically robust basis. However, for smaller water supply systems, particularly those serving less than 1000 people, where resources are likely to be very limited, it is reasonable to select plumbosolvency control conditions on a more generic basis, with reference to the typical conditions identified in this Code of Practice.

For the alleviation of iron discolouration, silicate doses of up to about 50 mg/l (as SiO_2) and polyphosphate doses of up to about 1 mg/l (P) have been used. Doses will be dictated by the severity of iron corrosion and the iron concentrations experienced, whether or not pH adjustment is concurrently needed, and will usually need to be established by pilot trials. An effective monitoring programme will be needed in order to judge success. This might include routine random sampling from customers' taps, the examination of distribution system pipe cut-outs and the use of automatic filter samplers at strategic locations.

In optimising general cuprosolvency control, it is important to appreciate the circumstances where copper leaching is a problem. Very long lengths of copper pipe-work in an institutional building and prolonged water stagnation might be best dealt with by changing the pipe-work configuration to reduce stagnation or by routine flushing, as opposed to the system-wide dosing of orthophosphate.

Pitting corrosion of copper pipe-work will require pipe replacement at the locations affected and the rectification of causative factors, such as low pH, aggressive disinfectant and chloride concentrations, or poor quality pipes.

Protocols for achieving optimised corrosion control for copper, iron, nickel and zinc are provided in Part B, Appendix 6.

3.5 PIPE REPLACEMENT

The need to rehabilitate old cast-iron water mains remains a major problem in many cities and towns and will have a bearing on the levels of service being provided to consumers in terms of "red water" discolouration, the frequency of bursts and areas of low pressure. Either the corroded mains are replaced, most commonly with medium density polyethylene (MDPE), or are scraped and relined with cement, epoxy resin or polymers. The alleviation of corrosion problems using silicates or polyphosphates should only be a short-term expedient.

Problems with long lengths of galvanised iron, copper or lead pipe-work within institutional buildings can result in high metal concentrations at the point of use. The elevated metals can be both direct corrosion by-products from the pipe material itself, but can also originate upstream of the premise plumbing sections and be accumulated within the corrosion deposit layer, to then be emitted intermittently at high levels thereafter. Part B, Appendix 7 provides guidance on the optimum design of pipe-work in institutional buildings, although opportunities for retrospective measures may be limited because of building layout and cost.

The principal long-term solution for reducing lead in drinking water will be the total removal of lead connection pipes and any remaining lead pipes within premises. However, the costs are high (between 1000 and 4000 Euros per house) and pipe replacement is both disruptive and inconvenient to consumers. The replacement of lead connection pipes is also complicated by split ownership between the water supplier and consumer. Generally, consumers have been reluctant to replace their section of lead pipe and most lead pipe replacement has been limited to the section owned by the water supplier (IWA, 2010; DWI, 2010).

There is ample evidence from case studies (DWI, 2010; IWA, 2010) that the partial replacement of lead connection pipes is likely to be insufficient, on its own, for achieving regulatory compliance with lead standards (also see Part B, Appendix 8). Indeed, partial lead pipe replacement can make matters worse, at least in the short-term, due to the disturbance of corrosion deposits and the possibility of metallic lead particles from cutting. If a section of lead pipe is replaced by copper, galvanic corrosion problems may be experienced if the copper is joined directly to the remaining section of lead piping. A plastic insert between the copper and lead should avoid such problems.

When the water supplier's section of lead connection pipe is replaced, every effort should be made to persuade the home-owner to replace their section of lead connection pipe at the same time. The full replacement of lead pipes effectively removes the lead exposure source without adverse spikes of lead release. Whenever lead connection pipes remain, whether in total or after partial replacement, a reservoir of high potential lead contamination remains. The development of new technologies for the lining of lead service pipes should be kept under review.

3.6 CONTROLLING THE USE OF METAL MATERIALS

In most countries, the use of lead piping for water supply is now prohibited, including its use in repairs to existing lead pipe-work. Solder containing lead is also prohibited for jointing copper pipe-work. Brass fittings should not contain lead greater than 0.25%.

Water suppliers should liaise with builders, plumbers and building managers in their area to check periodically that no inappropriate materials are being used. Occasional testing could also be considered.

3.7 CONTROLLING THE RATIO OF WATER FROM DIFFERENT SOURCES

When the water source is changed, the quality, including pH, alkalinity, sulphate, chloride and oxygen concentration, of the distributed water changes. A significant variation of water quality may induce the release of iron from the corrosion tubercles in old cast-iron pipes. Although adjusting pH or dosing orthophosphate or polyphosphate is effective for controlling iron release, the chemicals have a cost burden, which may be not affordable in poor countries. Thus, it is proposed to increase the ratio of water from the new source stage by stage rather than replace the water from the old source completely with that from the new source overnight (Gu, 2010). The optimal ratio of "new water" to "old water" can be determined experimentally. After the corrosion tubercles become stabilized with the "mixed water", the ratio can be adjusted. With this method, the iron release problem can be effectively controlled and no chemical is needed.

Additional reference

Gu Junnong (2010). Release of corrosion products in the old cast-iron pipes in case of multi-water sources and its control. *City and Town Water Supply*, **1**, 66–70.

Chapter 4

Implementing a monitoring programme

The monitoring programme should be capable of demonstrating that corrosion control objectives have been achieved and whether or not further measures or further optimisation are needed. It may comprise compliance monitoring to satisfy legal requirements, operational monitoring to better identify control needs and the performance of control measures, as well as supplementary investigations. A more comprehensive approach will provide a better basis for decision-making.

Compliance monitoring represents the bare minimum and may not be sufficient for informing corrosion control decisions (Hoekstra *et al.* 2008). Regulatory sampling frequencies are generally too low to gain a reliable conclusion because metal concentrations tend to be highly variable. In the US and Canada, compliance with copper and lead standards is based on stagnation samples, but results for lead are prone to distortion from water stood in non-lead pipe-work (IWA, 2010) and the targeting of copper sampling is often contrary to vulnerability to cuprosolvency effects. In the UK and the Netherlands, compliance is based on random daytime sampling, but sample numbers must be fairly high to overcome reproducibility problems.

Elsewhere in Europe, compliance monitoring for copper, lead and nickel varies, and in some countries may not be adequate, due to the failure of EU Member States to agree on a harmonised monitoring method. Recommendations were made to the European Commission in 2008 to adopt random daytime sampling for assessing the compliance of these metals (Hoekstra *et al.* 2008).

Operational monitoring should seek to overcome the limitations of compliance monitoring. It could involve (see Part B, Appendix 1):

- Sequential sampling after a period of stagnation;
- Increasing the frequency of random sampling; or
- Supplementing stagnation sampling with random sampling, the latter providing a system-wide perspective.

Investigations within a water supply system can provide further information on the extent of iron corrosion and may include the examination of pipe cut-outs, the use of automatic filter samplers, examining hydrant washings, and the use of coupons to measure corrosion rates. Examination of lead pipes removed from a system can indicate if Pb(II) or Pb(IV) chemistry is governing solubility (Part B, Appendix 2). Investigations at individual premises for lead can involve either split-flow or sequential stagnation sampling, as well as blood surveillance.

Chapter 5

Risk assessment

Risk assessment should be undertaken periodically and reviewed annually. The elements given in Tables 2.1 to 2.3 can form the basis for risk assessment in relation to controlling internal corrosion:

- Table 2.1. Diagnostic assessment
- Table 2.2. Circumstances associated with corrosion problems
- Table 2.3. Direct evidence of internal corrosion problems

Water suppliers will need to consider the wider implications of implementing a corrosion control measure and the possible significance of any planned changes to water treatment. Where phosphates are used as a corrosion inhibitor, the impact on sewage treatment in the area should be assessed as well as any environmental impact on receiving waters. This can be done by constructing a simple mass balance of all phosphate inputs and outputs.

For lead, water suppliers should consult all relevant stake-holders in their risk assessment, particularly health authorities, local medical practitioners and community representatives.

Part C provides a series of check-lists and criteria that can be used in risk assessment.

Chapter 6

On-going management

Internal corrosion control plans (see Section 3.1) should be reviewed on an annual basis, alongside risk assessments which should also be reviewed annually.

Liaison with stake-holders might include topic groups that focus on particular issues, such as lead pipe replacement to encourage consumer participation.

Communication with consumers should seek to explain how internal corrosion control problems are being tackled and give appropriate advice.

Whereas the emphasis of water suppliers will be their supplies and their customers, the appropriate local agencies should ensure that privately owned or non-municipal water supplies receive due attention, particularly in relation to lead because of the health implications. The problems particularly associated with small supplies are discussed in Part B, Appendix 9.

Effective corrosion control relies upon proper and consistent water chemistry within the appropriate target ranges. Therefore, the operational control of treatment processes and the management of water distribution systems need on-going evaluation and fine-tuning when necessary. Temporal and spatial fluctuations in water quality within a water supply system can readily undermine the effectiveness of a corrosion control programme.

Corrosion control systems may need to be adjusted in response to seasonal or longer term changes in climatic conditions, particularly temperature. Changes in the coagulant used in water treatment, particularly from metal sulphates (e.g. alum) to metal chlorides (e.g. PAC), may alter the chloride to sulphate mass ratio (CSMR) with the potential consequence of greater galvanic corrosion effects. The impact of seasonal increases in chloride from road salt should also be considered.

Chapter 7

Key references

The following key references relate to Part A, the Code of Practice. Additional references are provided, specific to the Foreword and to the Appendices in Part B, Supporting Information:

American Water Works Association (2011). Internal corrosion control in water distribution systems. AWWA Manual M58, First Edition.

AWWARF-TZW (1996). Internal corrosion of water distribution systems. 2nd Edition. AWWARF/DVGW–TZW, Denver, CO.

Drinking Water Inspectorate (2010). Guidance document. Guidance on the implementation of the Water Supply (Water Quality) Regulations 2000 (as amended) in England. September 2010.

Health Canada (2009). Guidance on controlling corrosion in drinking water distribution systems. Federal-Provincial-Territorial Committee on Health and the Environment. Ottawa, Ontario. June 2009.

Hoekstra E. J., Aertgeerts R., Bonadonna L., Cortvriend J., Drury D., Goossens R., Jiggins P., Lucentini L., Mendel B., Rasmussen S., Tsvetanova Z., Versteegh A. and Weil M. (2008). The advice of the Ad-Hoc Working Group on Sampling and Monitoring to the Standing Committee on Drinking Water concerning sampling and monitoring for the revision of the Council Directive 98/83/EC. European Commission Joint Research Centre, EUR 23374 EN – 2008.

International Water Association (2010). Best Practice Guide on the Control of Lead in Drinking Water. ISBN 10 1843393697

International Water Association (2012). Best Practice Guide on Sampling and Monitoring of Metals in Drinking Water. ISBN 978 1843393832

US Environmental Protection Agency (1991). Lead Copper Rule. *Federal Register*, **56**, (110), 26460–26564

Part B

Supporting Information

Appendix 1

Sampling methods and regulatory compliance

A1.1 SAMPLING METHODS USED IN COMPLIANCE ASSESSMENT

Regulatory standards for metals in drinking water are either based on aesthetic or health considerations or both. Whilst standards do differ across the world, both in terms of numeric values and the methods of compliance assessment, there are many similarities with World Health Organisation guidelines (WHO, 2011). For a selection of countries, the standards that apply for the metals of interest to this Code of Practice and pH are summarised in Table A1.1.

Table A1.1 Standards for metals (µg/l) and pH in drinking water.

Parameter	WHO	EU	US	Australia	Singapore	Canada
Copper	2000	2000	1300[1] 1000[2]	2000 1000	2000	1000
Iron	300[3]	200	300	300	–	300
Lead	10	25 and 10[4]	15[1]	10	10	15[1]or 10[5]
Nickel	70	20	–	20	70	–
Zinc	–	–	5000[2]	3000	–	5000
pH	6.5 to 8.5[3]	6.5 to 9.5[4]	6.5 to 8.5[2]	6.5 to 8.5	–	6.5 to 8.5[6]

[1] Action Level as 90th percentile, after 6+ hours stagnation

[2] National Secondary Standards (also stated that water should be "noncorrosive"). It is planned to remove these standards for pH following the revision of the US Lead Copper Rule

[3] WHO reference but not Guideline Value

[4] EU Directive 98/83/EC: 25 µg/l is the current Pb standard, 10 µg/l is the Pb standard from December 2013 (also stated for pH standards that "water should not be aggressive")

[5] Action Level after 30 minutes stagnation; action required if exceeded at >10% of sites sampled

[6] Operational guideline

Compliance with the US Lead Copper Rule (US EPA, 1991) is based on first draw one litre samples after at least 6 hours stagnation. It is common for pipework to be flushed before the stagnation period. For lead, the results from stagnation sampling are prone to distortion by dilution from water stood in non-lead pipe-work adjacent to the faucet (tap). Sequential sampling after the first draw sample should enable this distortion effect to be better appreciated, although laminar flow influences have been shown (Hayes & Croft, 2012) to skew lead emissions and reduce peak concentrations. US National Secondary Standards, such as those relating to pH, are not subject to enforcement unless adopted by State legislation. In Canada, compliance with guidelines for lead can either be based on sequential 30 minutes stagnation (30MS) samples and a standard of 10 µg/l or on sequential 6+ hours stagnation sampling and a standard of 15 µg/l (Health Canada, 2009).

In the UK, compliance for copper, lead and nickel is based on random daytime (RDT) sampling (DWI, 2010); one litre, first draw samples are taken (without flushing) at a random time during the day from houses selected at random. The Netherlands also bases compliance on RDT sampling. To overcome reproducibility problems, it is preferable that at least 100 samples (and if possible at least 200) should be taken each year from each water supply system, and it is beneficial to aggregate data over several years if operating conditions have remained constant. Recommendations have been made to the European Commission to adopt RDT sampling for assessing compliance with the EU standards for copper, lead and nickel (Hoekstra *et al.* 2008). In Singapore, copper is the major concern as there are no lead pipes in use and iron pipes are all cement-lined. Most sampling is undertaken in the water distribution network on a biannual basis and additional samples are collected from customer taps and supply mains when needed. In current practice, samples are collected after brief flushing. For specific investigation, pre-flush samples are collected.

The issues to be considered with the various sampling methods used in compliance assessment are summarised in Table A1.2 and further details can be obtained from IWA's Best Practice Guide on the Sampling and Monitoring of Metals in Drinking Water (IWA, 2012). In optimising internal corrosion control systems, it is recommended that more than one sampling method is used, if practicable.

Table A1.2 Issues associated with sampling methods.

Sampling method	Issues
Samples taken after flushing (typically for 2 to 5 minutes)	Does not characterise soluble metal pick-up from premise plumbing (Cu, Ni, Pb) but can indicate if there are problems in the distribution network (Fe)

(Continued)

Table A1.2 Issues associated with sampling methods *(Continued)*.

Sampling method	Issues
Samples taken during flushing	May indicate problems with particulate metals (Fe, Pb, Zn)
First draw sampling after stagnation (30 minutes or 6+ hours) for Cu & Pb	Samples are unlikely to contain water that has stood in lead service lines and may under-estimate the true extent of lead problems. Survey results will be variable due to on-going changes to the sampling pool and seasonal effects. Quality assurance may vary when consumers undertake the sampling. Surveys are not a sound basis for the optimisation of corrosion control.
Routine 30 minutes stagnation sampling at selected premises for Pb	Monthly sampling over several years can provide good performance data for optimisation purposes, but is sensitive to collaborating consumers withdrawing from the sampling programme. Pb concentrations must be high enough to be able to demonstrate change.
Sequential sampling after a stagnation period for Pb	Will provide a much better appreciation of lead problems at an individual property but profiles can be distorted by laminar flow effects.
Split-flow composite sampling – all metals	The most direct way of monitoring average concentrations over a selected period of time but logistics constrain widespread use.
Random daytime sampling – all metals	Provides an unbiased assessment of the extent of problems across a City or Town but requires a sufficiently large number of samples (at least 100) to overcome reproducibility problems. Not appropriate for characterising metal emissions at individual premises.

A1.2 SAMPLING METHODS USED AT INDIVIDUAL PREMISES IN RISK ASSESSMENT

Regulatory compliance tends to focus on the position across an entire water supply system. Assessing copper, lead or nickel emissions at an individual property will be best achieved by sequential stagnation sampling (30MS will be adequate for lead) or by split-flow composite sampling (IWA, 2012).

A1.3 SEQUENTIAL SAMPLING OF PREMISE PLUMBING

Sequential one-litre (or sub-litre) samples, taken after the stagnation period, can provide an indication of the profile of metal release in relation to plumbing characteristics from the water main up to the faucet (tap). The example given in Figure A1.1 shows peaks of lead associated with lead service lines, particular to each of the houses sampled, after 30 minutes stagnation.

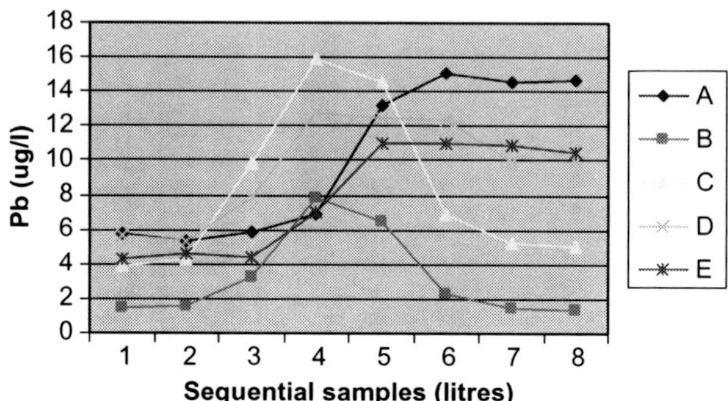

Figure A1.1 Sequential sampling results for lead after 30 minutes stagnation ** from Hayes and Croft (2012).

Field observations and modelling suggest that laminar flow influences can have a tendency to skew the lead results obtained, so that lower lead concentrations are observed over more samples than would be the case for plug flow, an approximation of the turbulent flow that would be expected in small diameter pipes under normal home use. Whether flow is laminar or turbulent will depend on flow rate, pipe geometry and pipe condition. Flows associated with normal tap use may differ from flows during sequential sampling.

In Figure A1.2, predicted results for sequential samples under plug and laminar flow conditions can be compared to those observed. The predicted results derive from the modelling of a home with a lead service line and copper premise plumbing (Hayes & Croft, 2012). In practice, the flow characteristics are likely to be somewhere in between the plug and laminar flow conditions that were assumed.

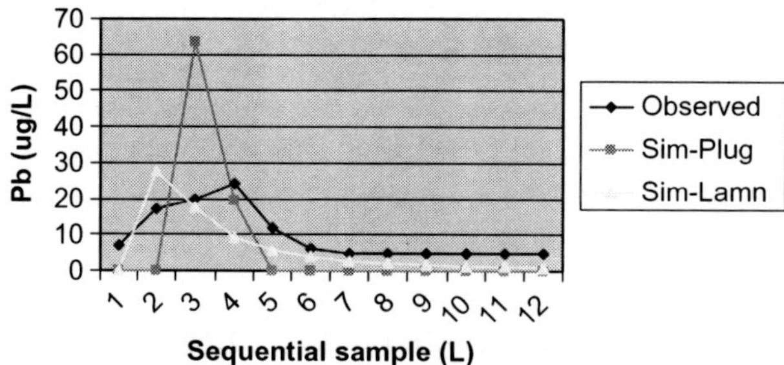

Figure A1.2 Predicted and observed sequential sampling results for lead after 6+ hours stagnation ** from Hayes and Croft, 2012.

For the lead concentrations at the tap (faucet), the skewing effects associated with sequential sampling have been shown by modelling (Hayes & Croft, 2012) to vary as a function of lead service pipe length and diameter and as a function of non-lead premise pipe length and diameter. Therefore, the results obtained from sequential sampling can only be used in a semi-quantitative manner, albeit repeated exercises from the same house will be helpful in the optimisation of plumbosolvency control.

A1.4 SUPPLEMENTARY OPERATIONAL MONITORING

It is important to appreciate that the minimum monitoring specified by regulations for compliance assessment may be inadequate for operational optimisation and control (Hoekstra *et al* 2008). Operational monitoring may comprise increased sampling frequency or the use of additional sampling methods, or both. Risk analysis should identify the extent of operational monitoring that is justified.

Additionally, on-line monitoring of pH and orthophosphate at treatment works should be considered. Such monitoring can be linked to alarm systems to ensure that any faults are rapidly discovered and resolved.

Additional reference

Hayes and Croft (2012). Collaborative project to determine proof of concept. Optimisation of corrosion control for lead in drinking water using computational modelling techniques. International Water Association, Specialist Group on Metals and Related Substances in Drinking Water, Research Report Series, IWA Publishing, London (in press).

Appendix 2

Corrosion testing

A2.1 TESTING OF METALLIC MATERIALS

A European Standard specifies a procedure to determine the release of metals from metallic materials used in construction products intended to come into contact with drinking water.

It involves a dynamic rig test for three purposes:

    (i)   Assessing a material as a reference for a category of materials using a range of water types with different compositions;

    (ii) Assessing a material for approval by way of comparative testing;

    (iii) Obtaining data on the interaction of local water with a material.

Additional reference

EN 15664-1:2008. Influence of metallic materials on water intended for human consumption – Dynamic rig test for assessment of material release – Part 1: Design and operation.

A2.2 LABORATORY PLUMBOSOLVENCY TESTING

The testing method that has been used extensively in the UK is based on the work of Colling *et al.* (1987). It involves pumping a batch of 50 litre samples through duplicate (in parallel) 15 cm lengths of new 12 mm internal diameter lead piping, at a flow rate (approximately 0.5 ml/min) to provide 30 minutes contact (30MC) between the test water and the piping. Each section of lead piping, sealed with rubber bungs, is immersed in a water bath and maintained at a constant 25°C for a test period of typically 20 days.

During the test period, the flow through each section of lead piping is regularly checked, along with pH of the test water, and samples of the test water leaving the lead pipe section are taken for lead analysis every few days. At these times, pH adjustments are made if necessary using small additions of acid or alkali. At the end of the test period, the test water is allowed to stagnate for 16 hours within the lead pipe section, prior to a further sample being obtained for lead analysis. Lead is then analysed using an appropriate method, which must be subject to routine quality control procedures.

A 32-channel peristaltic pump can be used, enabling 15 duplicate test samples and a duplicate control sample to be tested simultaneously. This allows various experimental designs to be used in investigating the plumbosolvency characteristics of test waters, examples being:

> (i)   15 different test waters as part of a regional survey, or
> (ii)   Three test waters, each at a particular pH, for five levels of ortho-phosphate addition, or
> (iii) A single test water over three pH conditions for five levels of ortho-phosphate addition.

Work by Colling *et al.* (1992) indicated that the lead emissions in equilibrium with sections of new lead pipe were similar to sections of old exhumed lead pipe, except that in the latter case the time taken for equilibrium to become established was very much longer (months as opposed to days with new lead pipe). They also found that lead emission concentrations approximately halved if the test temperature was reduced from 25 to 12°C, this relationship enabling test results to be extrapolated to average water supply temperatures. Further details can be obtained from the IWA's Best Practice Guide on the Control of Lead in Drinking Water (IWA, 2010).

Additional references

Colling J. H., Whincup P. A. E. and Hayes C. R. (1987). The measurement of plumbosolvency propensity to guide the control of lead in tapwaters. Journal of the Institution of Water and Environmental Management, Vol. 1, No. 3, December 1987.

Colling J. H., Croll B. T., Whincup P. A. E. and Harward C. (1992). Plumbosolvency effects and control in hard waters. JIWEM, Vol. 6, 259–268, June 1992.

A2.3 LEAD PIPE TEST RIGS

Lead pipe test rigs have many possible designs, including closed-loop and pass-through, the latter being more common. The lead pipe can be a straight length, a coil or swan-necked. In the UK, the swan-necked design has been the most popular and an automated example is shown in Figure A2.1. Other examples are given in AWWA (2011). In the lead pipe test rig shown, the operating principles are: (1) Water is flushed to waste; (2) Water is then held for a stagnation period of 30 minutes (or any other period); (3) The stagnated water is then captured by the sampling vessel which displaces earlier sample water to waste; and (4) The cycle is then repeated so that a fresh stagnation sample is available to be picked up at any time.

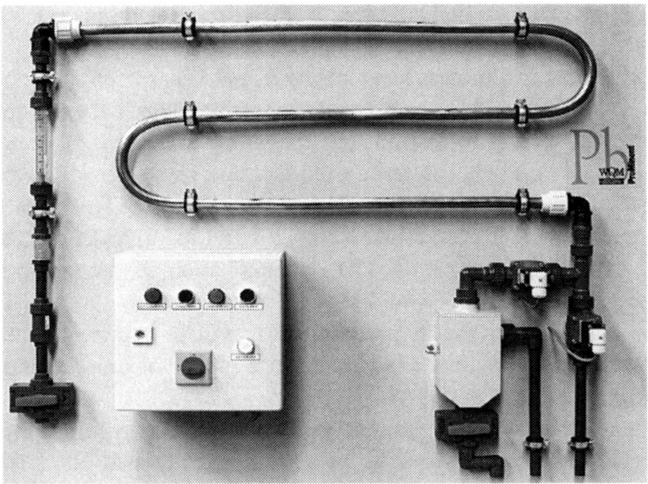

Figure A2.1 Example of a lead pipe test rig (from IWA, 2010).

New lead pipes were mostly used in the UK because of the erratic lead emissions that often derive from exhumed lead pipes. Test rigs can be deployed to monitor treated water at the treatment works, before and after ortho-phosphate dosing, and at strategic locations within the distribution network.

Additional references

American Water Works Association (2011). Internal corrosion control in water distribution systems. AWWA Manual M58First Edition .
AWWARF (1990). Lead control strategies. AWWARF, Denver, CO.
AWWARF (1994). Development of a pipe loop protocol for lead control. AWWARF, Denver, CO.

A2.4 THE NATURE OF PB(II) AND PB(IV) LEAD CORROSION DEPOSITS AND THEIR RELEVANCE TO SOLUBILITY

Tetravalent lead corrosion product, PbO_2, can be formed from the chlorination of lead-containing plumbing materials in the distribution system (Edwards & Dudi, 2004; Lytle & Schock, 2005; Schock & Lytle, 2010). In systems with lead service lines, metallic lead is first oxidized to divalent lead corrosion products such as $PbCO_3$ and $Pb_3(OH)_2(CO_3)_2$ then to PbO_2 (Liu *et al.* 2008). In systems using chloramines as the disinfectant, PbO_2 does not form (Switzer *et al.* 2006). The use of corrosion inhibitor such as orthophosphate may also inhibit its formation (Lytle *et al.* 2009). The size of PbO_2 particle is less than 100 nm and nanoparticle aggregates are typically found when they are present in the distribution system (Dryer & Korshin, 2007).

Free chlorine concentration as low as 1 mg/L as Cl_2 has been shown to be able to oxidize Pb(II) solids to PbO_2 and the 1:1 stoichiometry between free chlorine consumption and PbO_2 formation has been observed in laboratory studies (Zhang & Lin, 2011). PbO_2 is a strong oxidant and its stability is affected by the oxidation potential of the water. It has been found that PbO_2 can be reduced by natural organic matter (Dryer & Korshin, 2007; Lin & Valentine, 2008a) and Fe^{2+}/Mn^{2+} (Shi & Stone, 2009). The decomposition of monochloramine may also cause reductive dissolution of PbO_2 (Lin & Valentine, 2008b, 2009). Field observations have shown that a predominantly Pb(II) hydroxyl-carbonate scale can be converted into a mainly PbO_2 scale in a matter of months under high pH conditions (Schock & Lytle, 2010).

Thus, changes in water treatment processes that may alter the oxidation potential of drinking water need to be carefully evaluated for lead release if the system has lead service lines and a history of using free chlorine as the disinfectant.

To put the above into perspective, XRD analysis by the US EPA of lead pipe sections from 48 water supply systems revealed that 29 (60.4%) were dominated by Pb(II) corrosion deposits, 13 (27.1%) were partially dominated by Pb(II) corrosion deposits, and 6 (12.5%) were dominated by Pb(IV) corrosion deposits. Water utilities should undertake the analysis of a small number (say, 6) of exhumed lead pipe sections from representative locations in their supply area to determine which lead corrosion compounds are dominant, as this is relevant to the optimisation of plumbosolvency control.

Additional references

Dryer D. J. and Korshin G. V. (2007). Investigation of the reduction of lead dioxide by natural organic matter. *Environ. Sci. Technol.*, **41**(15), 5510–5514.

Edwards M. and Dudi A. (2004). Role of chlorine and chloramine in corrosion of lead-bearing plumbing materials. *J. Am. Water Works Assoc.*, **96**(10), 69–81.

Lin Y. P. and Valentine R. L. (2008a). The release of lead from the reduction of lead oxide (PbO_2) by natural organic matter. *Environ. Sci. Technol.*, **42**(3), 760–765.

Lin Y. P. and Valentine R. L. (2008b). Release of Pb(II) from monochloramine mediated dissolution of lead oxide (PbO_2). *Environ. Sci. Technol.*, **42**(24), 9137–9143.

Lin Y. P. and Valentine R. L. (2009). Reduction of lead oxide (PbO_2) and release of Pb(II) in mixtures of natural organic matter, free chlorine and monochloramine. *Environ. Sci. Technol.*, **43**(10), 3872–3877.

Liu H. Z., Korshin G. V. and Ferguson J. F. (2008). Investigation of the kinetics and mechanisms of the oxidation of cerussite and hydrocerussite by chlorine. *Environ. Sci. Technol.*, **42**(9), 3241–3247.

Lytle D. A. and Schock M. R. (2005). Formation of Pb(IV) oxides in chlorinated water. *J. Am. Water Works Assoc.*, **97**(11), 102–114.

Lytle D. A., Schock M. R. and Scheckel K. (2009). The Inhibition of Pb(IV) Oxide Formation in Chlorinated Water by Orthophosphate. *Environ. Sci. Technol.*, **43**(17), 6624–6631.

Shi Z. and Stone A. T. (2009). PbO2(s, Plattnerite) Reductive Dissolution by Aqueous Manganous and Ferrous Ions. *Environ. Sci. Technol.*, **43**(10), 3596–3603.

Schock M. R. and Lytle D. A. (2010). Internal corrosion and deposition control, Water Quality and Treatment: A Handbook for Community Water Supplies. 6th Edition. McGraw–Hill inc, New York.

Switzer J. A., Rajasekharan V. V., Boonsalee S., Kulp E. A. and Bohannan E. W. (2006). Evidence that monochloramine disinfectant could lead to elevated Pb levels in drinking water. *Environ. Sci. Technol.*, **40**(10), 3384–3387.

Zhang Y. and Lin Y. P. (2011). Determination of PbO_2 formation kinetics from the chlorination of Pb(II) carbonate solids via direct PbO_2 measurement. *Environ. Sci. Technol.*, **45**(6), 2338–2344.

A2.5 GALVANIC CORROSION TESTING

Bench-scale and pilot-scale studies that evaluate lead release into drinking water do not typically incorporate galvanic cells between leaded plumbing materials and copper pipe, even though such configurations are often found in premise plumbing and in public distribution systems.

Bench-Scale Test

A simple bench-scale test was recently developed in the US, which can allow drinking water utilities to assess lead leaching to water from galvanic connections of leaded solder to copper pipe. This simple dump-and-fill protocol was developed to rapidly screen for significant changes in lead leaching that could result from various water treatment changes by drinking water utilities (Nguyen *et al.* 2010). The bench-scale test uses a 1-inch long copper coupling (½-inch diameter copper) with a 1-inch long 50:50 Pb/Sn solder wire melted inside. The solder-copper coupling is then exposed to 100 mL of test water inside a glass vial. Three replicates are tested for each water condition to ensure statistical confidence in key trends. Water exposed to the simulated copper joint inside each vial is changed twice per week and it is otherwise stagnant. Composite weekly samples are routinely collected from each vial and the unfiltered composite samples are then analyzed for metals. All materials are kept at room temperature throughout the testing period. It should be noted that the lead levels released from "fresh" solder under such an extreme water stagnation scenario are considered "worst case", and they markedly overestimate field levels of lead. At the same time, the bench-scale test is qualitatively useful in comparing trends in lead leaching from the simulated solder joints, when a drinking water utility is considering implementing a water change with unknown consequences to water corrosivity. In most cases, short-term tendencies as measured by the test were in qualitative agreement with practical experiences of ten US drinking water utilities and with longer-term test results (Nguyen *et al.* 2010).

Pilot-Scale Test

A portable pilot test rig was also recently developed in the US, as part of a Water Research Foundation project entitled "Non-Intrusive Methodology for Assessing Lead and Copper Corrosion". The rig will allow water utilities to monitor trends in lead and copper leaching from representative premise plumbing materials, without involving consumers and by using standardized metallic rigs (Cuppett, 2012). The proposed preliminary design includes a lightweight, yet sturdy, support structure; plastic (primarily PVC) wetted surfaces aside from the test metals; and an easy to operate timer for programming cyclical flow and stagnant water conditions. The pilot test rig includes triplicate lead pipe: copper pipe and leaded solder: copper pipe galvanic couples with the ability to disconnect, and is currently being tested by five drinking water utilities in the US.

Preceding this portable pilot-scale rig is another pilot-scale test which examines lead leaching from four representative configurations of service lines including: 1) 100% lead (Pb), 2) 100% copper (Cu), 3) 50% Pb upstream of 50% Cu to assess galvanic corrosion, and 4) 50% Pb downstream of Cu to assess galvanic and deposition corrosion, using a range of flow rates (Cartier *et al.* 2012).

Additional references

Cantor A. F. (2009). Water Distribution System Monitoring. CRC Press, Taylor & Francis Group, Boca Raton, FL.

Cartier C., Arnold R., Triantafyllidou S., Prevost M. and Edwards M. Long-term Effect of Flow Rate and Lead/Copper Pipe Sequence on Lead Release from Service Lines. *Water Research*, DOI: 10.1016/j.watres.2012.05.010, 2012

Cuppett J. (2012). Lead and Copper Rule and Distribution System Corrosion: An Overview of Foundation Research. Water Research Foundation report.

Nguyen C., Stone K., Clark B., Edwards M., Gagnon G. and Knowles A. (2010). Impact of Chloride: Sulfate Mass Ratio (CSMR) Changes on Lead Leaching in Potable Water. Final Report, Water Research Foundation, Denver, CO.

Appendix 3

Compliance modelling

The models enable the most relevant features of a water supply zone to be incorporated in the prediction of zonal compliance with lead standards, as a function of both plumbosolvency (corrosivity of the water to lead) and the zone's physical characteristics. Compliance can be predicted for various sampling protocols. Additionally, the models can investigate lead emissions from individual premises.

The zonal model simulates the emissions of lead at individual simulated houses, through time, across an entire water supply area such as a City or Town. It uses a single pipe model to determine the lead emission profile at each simulated house, the characteristics of each simulated house being the outcome of the random ascription of variables, which follows the Monte Carlo method for establishing a probabilistic frame-work. The single pipe model simulates the dissolution of lead into the water flowing through or stagnating in a coupled lead pipe and non-lead pipe, over a 24-hour period. Particulate lead is not simulated but can be investigated indirectly. The model does not simulate lead release from brass or galvanic corrosion. The coupled pipes are broken down into a series of "elements" and when assuming simple plug flow, each element is treated as a stirred tank, flow being simulated by passing the contents of each stirred tank to the next at a time interval of one second.

The rate of lead dissolution is determined by reference to an exponential curve that declines towards equilibrium. As M (the initial mass transfer rate which determines the initial slope of the dissolution curve) and E (the equilibrium concentration) reduce, the water is less plumbosolvent (less lead dissolves over time). These factors can be determined by stagnation sampling at appropriate reference houses or by laboratory plumbosolvency testing. Either plug flow (an approximation of turbulent flow) or laminar flow can be modelled. When the imaginary tap is closed (that is, zero flow), the lead concentration increases over time as determined by M and E. When the imaginary tap is open, the concentration of lead in the emission from the pipe is either (i) a reflection of the steady state flushed condition (with lead concentrations normally below 1 µg/l unless the lead pipe is very long) or (ii) it is determined by previous zero flow (stagnation) conditions, as influenced by pipe geometry and the extent of the flow event.

The zonal model is set up by the random ascription of a series of zonal characteristics, as derived from sets of agreed statistical distributions, and by the use of agreed variables and constants. The aim of this probabilistic Monte Carlo framework is to describe the huge variation that undoubtedly occurs in real water supply zones. If we can mimic this real-world variation then the model can be used for predictive purposes, as has been demonstrated by numerous case studies (Hayes *et al.* 2006, 2008). By changing the plumbosolvency factors M and E, treatment conditions can be evaluated in terms of zonal compliance.

In addition to defining the plumbosolvency of the water supplies in numerical terms from laboratory testing, the following calibration data will be required:

- The statistical distribution of the lengths and diameters of lead service lines (including the part owned by the house-holder) – this is best determined by the on-going inspection of the houses that are sampled for regulatory purposes, or assumptions can be used with reference to circumstances elsewhere;
- The statistical distribution of the lengths and diameters of non-lead pipes (essentially premise plumbing)–this is best determined by the on-going inspection of the houses that are sampled for regulatory purposes, or assumptions can be used with reference to circumstances elsewhere;
- The statistical distribution of daily volumetric consumption – this is best determined from meter records or assumptions can be used with reference to circumstances elsewhere, but is only relevant to compliance based on random daytime sampling;
- The percentage of houses in the water supply system that have a lead service line – this is mainly relevant to compliance based on random daytime sampling;
- The pattern of water use - assumptions are used, but are only relevant to compliance based on random daytime sampling.

The level of calibration will depend on the overall availability of data and its accuracy. Commonly, assumptions will need to be made, at least for initial modelling. Validation is based on the comparison of predicted to observed survey results. Alternatively, the model can be calibrated by adjusting data input to fit the predicted survey results to those observed; in such cases, it is not possible to validate the model to the same extent, but this does not preclude using the model for predictive purposes. A major use of the model is to investigate different treatment scenarios in relation to compliance criteria and in so doing to be able to select optimum treatment conditions.

Further details of the modelling procedures can be obtained from IWA's Best Practice Guide for the Control of Lead in Drinking Water (IWA, 2010).

Additional references

Hayes C. R., Bates A. J., Jones L., Cuthill A. D., Van der Leer D. and Weatherill N. P. (2006). Optimisation Of Plumbosolvency Control Using A Computational model. *Water And Environment Journal* **(20)** 256–264.

Hayes C.R., Incledion S. and Balch M. (2008). Experience In Wales (UK) Of The Optimisation Of ortho-Phosphate Dosing For Controlling Lead In Drinking water. *Journal Of Water And Health.* **06.2**, 177–185.

Appendix 4

Definition of the term "optimisation" as it relates to the control of lead in drinking water

A4.1 BACKGROUND

The definition of the term "optimisation" has been very vague in all the guidance issued by regulators in Canada, the US and the UK. There is little or no guidance elsewhere.

In Canada, optimisation has been linked to compliance with standards based on sequential stagnation sampling, either after 6+ hours or 30 minutes stagnation. The precise numerical method for using sequential sampling results after 6+ hours stagnation is unclear, whereas the averaging of results after 30 minutes stagnation is open to distortion depending on pipe-work circumstances. Some Provinces have their own standard based on two sequential 30 minutes stagnation samples, whereas the national guideline is based on samples taken after flushing. The basis for optimising plumbosolvency control is therefore uncertain.

In the US, corrosion control is a regulatory requirement in cases of non-compliance with the Lead Copper Rule (LCR). The use of the first litre drawn from the faucet after 6+ hours stagnation is also open to distortion and downward bias, depending on pipe-work circumstances, and LCR surveys are prone to variation from on-going changes to the sampling pool, seasonal effects, variations in sampling protocol (such as flushing before stagnation) and from a reliance on consumer based sampling. In consequence, the requirement for implementing corrosion control may be unclear and vary purely by chance. The approach taken to corrosion control also differs from State to State. Again, the basis for optimising plumbosolvency control is uncertain.

In England and Wales, the Drinking Water Inspectorate issued guidance to water companies in 2000 (updated in 2001) that required them to optimise corrosion control as it related to lead in drinking water. This guidance stated, in cases where orthophosphate dosing was required (95% of all water supply systems operated by the water companies), that "optimisation means applying the optimum orthophosphate dose and maintaining the optimum orthophosphate concentration within distribution within the optimum pH range to obtain the best practical reduction in lead concentrations." The Inspectorate then indicated that the optimum dose of orthophosphate and "best practicable reductions in lead concentrations" could be determined from criteria based on:

- Laboratory plumbosolvency testing;
- Full scale or pilot scale trials;
- Practical experience elsewhere under similar circumstances;
- Solubility or compliance modelling;
- Random daytime sampling if no more than 2% exceeded 10 µg/l;
- Increases in orthophosphate doses producing no further worthwhile reduction in lead concentrations.

The optimisation programmes undertaken by the water companies were subject to legal agreement with the Secretary of State, were subject to annual audit and formal reporting requirements, and have so far achieved 99% compliance nationally with 10 µg/l based on random daytime sampling (with 99.5% compliance in some regions).

Terms such as "worthwhile reductions in lead" were not defined by the Inspectorate but were related to the practical experiences demonstrated by the water companies and as assimilated by the Inspectorate through audit.

A4.2 BEST AVAILABLE TECHNIQUES NOT ENTAILING EXCESSIVE COST (BATNEEC)

The UK approach outlined above follows the "BATNEEC" principle that has been widely and successfully used in Europe for industrial pollution control. In the context of corrosion control, to achieve the best practicable reductions in lead concentrations in drinking water:

- "best available" relates readily to well established treatment methods;
- "techniques" encompasses not only centralised treatment (albeit this is often the focus) but allied approaches such as lead pipe replacement, flushing, point-of-use treatment, educational programmes, and so on; this holistic definition should also extend to the actions taken by water utilities to minimise interference effects from natural organic matter and iron particulates;
- "not exceeding excessive cost" is readily applicable to the costs associated with pH control and orthophosphate dosing, when the latter has a typical unit operating cost of only 0.5 Euro-Cents per cubic meter of water supplied; however, there could be "excessive costs" associated with widespread rehabilitation of old cast iron water mains in some areas, whereby rehabilitation phasing might need to be considered in the on-going definition of "optimisation".

A4.3 A GENERIC DEFINITION OF THE TERM "OPTIMISATION" AS IT RELATES TO THE CONTROL OF LEAD IN DRINKING WATER

Optimisation of corrosion control as it relates to lead in drinking water can be taken to mean:

"The application of best available techniques, not exceeding excessive cost, to reduce lead concentrations in drinking water to the minimum that is practical to achieve".

A holistic approach is implied that is robust scientifically, that incorporates risk assessment and that matches mitigation measures specifically to the circumstances of the water supply system. In practice, the working definition of the term "optimisation" will be strongly influenced by the compliance methods, guidance and standards imposed by the regulatory agencies. In this context, three assumed compliance scenarios have been used in Appendix 5 (that follows) for defining the need for the optimisation of corrosion control treatment to minimise lead in drinking water.

Appendix 5

Protocols for the optimisation of corrosion control treatment to minimise lead in drinking water

INTRODUCTION

Three protocols are described that reflect the different circumstances of water supply systems, perceived priorities and the availability of resources:

(a) *Science based* – the most comprehensive approach that incorporates a range of state-of-the-art assessment techniques allied to confirmation by subsequent monitoring; likely to achieve optimisation earlier and with greater accuracy. Applicable to larger systems serving more than 100,000 people.

(b) *Trial based* – a simpler approach based on judgement from available knowledge and iterative assessment via monitoring: although initial costs could be lower, the possibly longer time taken to achieve optimisation and the greater potential for error could result in higher costs overall.

(c) *Generic* – applicable to smaller systems when resources are very limited.

A. SCIENCE BASED
Step 1 Confirm need for optimisation

Optimisation will be required, or should at least be investigated, if either:

(a) The 90th percentile concentration of lead exceeds the Action Limit of 15 µg/l in a survey (at a representative time of the year) based on at least four (and preferably twelve) sequential samples (each of one litre volume) taken after flushing and 6+ hours stagnation from a minimum of 50 houses, of which a minimum of 50% should have a lead service line; 90th percentiles should relate to each series of the 1st, 2nd, 3rd and 4th and so on, samples taken and action is required if the 90th percentile concentration in any sample series exceeds the Action Limit; or

(b)   More than 10% of sites have a concentration of lead that exceeds the Action Limit of 10 µg/l in a survey (at a representative time of the year) based on at least four (and preferably twelve) sequential samples (each of one litre volume) taken after flushing and 30 minutes stagnation from a minimum of 50 houses, all of which should have a lead service line; assessments should relate to each series of the 1st, 2nd, 3rd and 4th and so on, samples taken and action is required if more than 10% of samples in any sample series exceeds the Action Limit; or

(c)   More than 1% of at least 100 (and preferably at least 200) random daytime (first draw, one litre) samples exceed 10 µg/l, taken over an annual period. For small water supply systems, it may only be possible to take a smaller number of samples, in which case the results from several years can be aggregated if there have been no significant operational changes.

If the criteria in (a), (b) or (c) are not exceeded, then treatment can be considered to be optimised in general terms, unless risk assessment indicates otherwise. In cases where treatment is considered to be optimised, any corrective action is likely to be confined to dealing with problems that might arise at individual premises. Surveys should be undertaken routinely on at least an annual basis, with particular regard to any significant changes in water treatment.

Step 2 Investigate the nature of lead corrosion deposits

For water supply systems with high free chlorine residuals (generally, greater than 1 mg/l), if the need for optimisation is confirmed by Step 1, a minimum of 6 lead pipe cut-outs from lead service lines, taken from across the water supply system, should be examined by XRD analysis, to determine the nature of the lead corrosion deposits. If the corrosion deposits are dominated by Pb(II) compounds, proceed to Step 3. If the corrosion deposits are dominated by Pb(IV) compounds, optimisation of treatment may not be necessary, or may only achieve very marginal gains. In such cases, the emphasis should be to determine the causes of exceedance at individual premises and take appropriate corrective action. Step 3 will be inappropriate if the corrosion deposits are dominated by Pb(IV) compounds but Step 4 can be considered for gaining a deeper appreciation of regulatory compliance.

Step 3 Laboratory plumbosolvency testing

A representative sample of the treated water in supply should be tested to determine its plumbosolvency and its response to orthophosphate treatment under three pH conditions (see Appendix 2 for method summary). The pH conditions and orthophosphate concentrations that are evaluated will be influenced by the alkalinity of the water. The results should indicate the orthophosphate dose response of the water for an applicable range in pH and can be used in compliance modelling to determine optimum treatment conditions (Step 4).

Step 4 Compliance modelling to identify optimum treatment conditions

Compliance modelling (see Appendix 3 for method summary) can utilise the results from Step 3 to investigate the relationship between treatment conditions and the level of compliance likely to be achieved, for each survey type indicated in Step 1.

Step 5 Implementation and performance appraisal

As a prerequisite to Steps 1 to 4, it will be beneficial to undertake a diagnostic assessment of the water supply system, including the identification of any possible interference effects from natural organic matter or iron particulates. This and Steps 1 to 4 can readily be undertaken over a period of three months from which the likely optimum treatment conditions can be determined.

To confirm the performance of the optimum treatment conditions identified will best be achieved by using a combination of monitoring techniques, including:

- Compliance monitoring, as prescribed by the regulatory agencies, notwithstanding the limitations of the sampling methods involved;
- Sequential sampling after stagnation to provide an indication of the lead emission profile of lead service lines, notwithstanding possible skewing effects;
- Routine (at least monthly) performance monitoring at selected houses that have a lead service line, based on sequential sampling after 30 minutes stagnation – the lead emissions from the houses selected must be sufficiently high to be able to demonstrate the effect of the treatment change imposed; as an alternative to sequential sampling, a ten litre bulk sample could be taken after stagnation to investigate the general level of lead contamination associated with the home's pipework system;
- The deployment of lead pipe test rigs, using new lead piping, at strategic locations in the water supply system; by using new lead piping, responses to water quality changes will be rapid;
- The deployment of lead pipe test rigs, using old lead piping, at for example: (i) after water treatment but prior to the dosing of corrosion inhibitor, and (ii) at the water treatment plant after the dosing of corrosion inhibitor – responses to water quality changes can be slow, up to several years, and results can be erratic if pipe deposits have been disturbed during exhumation from service or in rig installation.

The more of these monitoring techniques that are used, the clearer will be the conclusions that can be achieved. There will be benefits in undertaking pilot trials if timescales permit. Consideration should also be given to the possible effects from metal precipitates within the water supply system.

Random daytime sampling, as a supplement to stagnation based surveys, will give a system-wide perspective of lead emissions and permit population exposure estimates to be determined. It will also provide an opportunity to check the extent of any attenuation of corrosion inhibitors, pH drift and changes in free chlorine that might occur across the water supply system. Consideration should be given to any part of a water supply system where pH might rise because of cement lining. Any differences in free-chlorine due to attenuation might vary the presence of Pb (IV) corrosion deposits as opposed to Pb(II). On-line monitoring of pH and orthophosphate at the water treatment plant is strongly recommended for checking the integrity of chemical dosing systems.

B. TRIAL BASED
Step 1 Confirm need for optimisation

Same as in Step 1 of Protocol A.

Step 2 Selection of anticipated optimum treatment conditions

The selection of anticipated optimum treatment conditions should follow this Code of Practice, taking into account local circumstances. The justification for the selection of the treatment conditions should be recorded.

Step 3 Implementation and performance appraisal

Confirmation of the performance of the optimum treatment conditions selected will best be achieved by using a combination of monitoring techniques, as listed in Protocol A. Reliance on compliance monitoring alone is not recommended. Dependent on results, treatment conditions may need to be adjusted.

C. GENERIC

The need for action should be based, at the minimum, on a risk assessment, recognising that compliance monitoring may be too limited. The risk assessment and any corrective actions taken should follow this Code of Practice as far as possible. The conclusions of risk assessment and the corrective actions taken should be recorded.

For smaller water systems, including building systems serving sensitive subpopulations such as children and expectant mothers, the best approach will be to start with an assessment of leaded materials throughout the entirety of the water distribution system, to each tap used for human water consumption. Where accessible, lead pipes or plumbing connections and devices composed of leaded solder or alloys should be removed and appropriate product and plumbing components consistent with the most updated manufacturing standards and plumbing codes should replace them.

There should also be conscientious and stringent application of the most appropriate international and local plumbing standards to prevent the future installation of components comprised of brasses containing leachable lead, or comprised of or containing other leachable metals of health concern. Local health authorities and water utilities should work closely together on such issues.

Appendix 6

Protocols for the optimisation of corrosion control for copper, iron, nickel and zinc in drinking water

6.1 COPPER
Step 1 Confirm need for optimisation

Optimisation will be required, or should at least be investigated, if either:

(a) The 90th percentile concentration of copper exceeds 1000 µg/l (or 1300 µg/l where applicable) in a survey, at a representative time of the year, based on first draw, one litre samples taken after flushing and 6+ hours stagnation from a minimum of 50 houses; or

(b) More than 10% of sites have a concentration of copper that exceeds 500 µg/l in a survey (at a representative time of the year) based on first draw, one litre samples taken after flushing and 30 minutes stagnation from a minimum of 50 houses; or

(c) More than 1% of at least 100 (and preferably at least 200) random daytime first draw, one litre samples exceed 1000 µg/l, taken over an annual period. For small water supply systems, it may only be possible to take fewer samples, in which case the results from several years can be aggregated if there have been no significant operational changes.

It has been assumed that monitoring for lead is being undertaken by one or more of the above approaches (a) to (c) and that all that is required is for copper to be analysed additionally.

If the criteria for copper in (a), (b) or (c) are not exceeded, then treatment can be considered to be optimised in general terms, unless risk assessment indicates otherwise. In cases where treatment is considered to be optimised, any corrective action is likely to be confined to dealing with problems that might arise at individual premises. Surveys should be undertaken routinely on at least an annual basis, with particular regard to any significant changes in water treatment.

The elements of risk assessment that should also be taken into account are:

- The number of complaints from consumers about aesthetic effects (e.g. green staining of sanitary ware);
- The occurrence of any copper pitting corrosion problems;
- The pH of the water supplies and its stability; and
- The number of new constructions in the system and the likelihood of cuprosolvency problems.

Additional evidence can be obtained from the examination of copper pipe cut-outs.

Step 2 Determine possible optimisation measures

For general cuprosolvency problems, the measures to be considered should include:

- pH adjustment;
- The selection of corrosion inhibitor; and
- The concentration of the corrosion inhibitor.

For copper pitting corrosion, unless widespread, each case should be thoroughly investigated to determine the cause and localised action taken as necessary.

The investigations should distinguish between residential and institutional buildings and the age of the pipework involved.

Step 3 Implementation and performance appraisal

Confirmation of the performance of the optimum treatment conditions selected will best be achieved by using a combination of monitoring techniques:

- 30 minutes stagnation sampling on a routine basis, recommended to be at least monthly from at least 6 homes;
- Random daytime sampling of the water supply system, with at least 50 samples per annum. For small water supply systems, it may only be possible to take fewer samples, in which case the results from several years can be aggregated if there have been no significant operational changes.

If at all possible, monitoring should be undertaken both before and after any changes to the corrosion control treatment.

If cuprosolvency problems are mainly confined to large institutional buildings with long-established premise plumbing (more than 2 years old), modification of pipework to reduce water stagnation may be warranted.

If cuprosolvency problems are mainly confined to newly installed copper pipes, and if such problems are considered to be significant, it may be necessary to restrict the use of copper piping. Any restrictions applied must be properly justified.

For water supply systems serving more than 100,000 people, cuprosolvency control should be considered thoroughly in the annual risk assessment of corrosion control. For smaller systems, a more limited and reactive approach may apply. In all cases, water utilities should endeavour to liaise with all the relevant parties involved, such as the local health authority and plumbing businesses.

6.2 IRON
Step 1 Confirm need for optimisation

It will first be necessary to determine the nature of any problems with iron in the water supply system:

- Some groundwaters naturally contain high concentrations of iron (up to several mg/l) – in such cases, oxidation and filtration treatment processes will be required (IWA, 2012b);
- In the treatment of surface derived waters, the use of iron-based coagulants can often result in iron breaking through into the supply system – optimisation of the coagulation process will be required (IWA, 2012b);
- Other than source related issues, the commonest problem with iron is due to the corrosion of old cast-iron water mains;
- Additionally, iron pick-up can occur with galvanised iron service pipes or premise plumbing.

It is these latter two sources of iron that are relevant to corrosion control.

Optimisation will be required, or should at least be investigated, if either:

(a) The 90th percentile concentration of iron exceeds 300 µg/l in a survey (at a representative time of the year) based on first draw, one litre samples taken after flushing and 6+ hours stagnation from a minimum of 50 houses; or

(b) More than 10% of sites have a concentration of iron that exceeds 100 µg/l in a survey (at a representative time of the year) based on first draw, one litre samples taken after flushing and 30 minutes stagnation from a minimum of 50 houses; or

(c) More than 1% of at least 100 (and preferably at least 200) random daytime first draw, one litre samples exceed 100 µg/l, taken over an annual period. For small water supply systems, it may only be possible to take fewer samples, in which case the results from several years can be aggregated if there have been no significant operational changes.

It has been assumed that monitoring for lead is being undertaken by one or more of the above approaches (a) to (c) and that all that is required is for iron to be analysed additionally.

If the criteria for iron in (a), (b) or (c) are not exceeded, then corrosion control can be considered to be optimised in general terms, unless risk assessment indicates otherwise. Surveys should be undertaken routinely on at least an annual basis, with particular regard to any significant changes in water treatment.

The elements of risk assessment that should also be taken into account are:

- The number of complaints from consumers about aesthetic effects (e.g. brown staining of washing or sanitary ware);
- The occurrence of "red-water" iron discolouration episodes and water main bursts, and the extent of low pressure problems in the system;
- Evidence of the accumulation of iron in pipework;
- The amount of leakage in the system; and
- The pH of the water supplies and its stability.

Additional evidence can be obtained from the examination of iron water main cut-outs, from the flushing of hydrants and by installing automated water filtration sampling devices at strategic locations.

Step 2 Determine possible optimisation measures

For general iron corrosion problems, the measures to be considered should include:

- pH adjustment;
- The selection of corrosion inhibitor;
- The concentration of the corrosion inhibitor;
- The replacement of old cast-iron water mains; and
- The refurbishment of old cast-iron water mains by scraping and re-lining.

Particularly for water supply systems serving more than 100,000 people, the water utility should catalogue the principal components of its water supply system, including the type (iron, cement, plastic, etc.) and size of all water mains greater than 50 mm diameter, storage points with capacities, the demand zones with the domestic populations served, and any significant industrial consumers. The water utility should use a network model to investigate corrective measures. Source related problems with iron should be resolved first, before the rehabilitation of water mains. As a short-term palliative, proprietary polyphosphates and silicates can be considered.

Due to the costs involved, the rehabilitation of water mains across a supply system will normally be phased. Whilst priorities for action can be determined from the investigations and risk assessments undertaken, it may be beneficial to undertake rehabilitation work first that is closest to the supply source and progress out towards the extremities.

Step 3 Implementation and performance appraisal

Confirmation of the performance of the optimum treatment conditions selected and the progress of a programme of mains rehabilitation will best be achieved by using a combination of monitoring techniques; these could include:

- Random daytime sampling of the water supply system for iron, with at least 50 samples per annum. For small water supply systems, it may only be possible to take fewer samples, in which case the results from several years can be aggregated if there have been no significant operational changes.

- Analysing iron in any samples taken after stagnation for other metals;
- The continued use of automated water filtration sampling devices at strategic locations and the flushing of hydrants; and
- Continuing to record the number of complaints from consumers about aesthetic effects, the occurrence of "red-water" iron discolouration episodes and water main bursts, and the extent of low pressure problems in the system.

For water supply systems serving more than 100,000 people, iron discolouration should be considered thoroughly in the annual risk assessment of corrosion control. For smaller systems, a more limited and reactive approach may apply. In all cases, water utilities should endeavour to liaise with all the relevant parties involved.

Additional reference

International Water Association (2012b). Best Practice Guide on Metals Removal from Drinking Water by Treatment. ISBN 9781843393849

6.3 NICKEL

Little attention has been given to nickel in drinking water. The principal source is from the chrome-nickel plated components of water fittings, particularly taps. No specific action is likely to be warranted although nickel leaching should at least be kept under review by periodic risk assessment.

It is suggested that nickel is analysed in any samples taken for metal analysis, although random daytime sampling can be used to determine the specific extent of nickel leaching across a water supply system. Circumstantial evidence from the UK suggests that orthophosphate inhibits the leaching of nickel.

6.4 ZINC

Little attention has been given to zinc in drinking water and any elevated zinc is only normally considered to be of aesthetic significance (WHO, 2011). A major source is brass in which zinc is a component and another is galvanised steel/iron pipes where used as service pipes or in domestic installations, especially those aged over 30 years. In mixed installations (e.g. galvanised steel/copper pipes or galvanised steel pipes/brass fittings) electrochemical corrosion can occur which may accelerate dezincification and dissolution of zinc layers from galvanised pipes.

It is suggested that zinc is analysed in any samples taken for metal analysis, although random daytime sampling can be used to determine the specific extent of zinc leaching across a water supply system. In risk assessment, zinc may be considered to warrant further consideration if the pH of water supplies exceeds 8.3 and the chloride to alkalinity ratio exceeds 0.5, or if there have been complaints from consumers of "sandy water" or opalescent appearance of water and/or greasy films on boiling. The presence of elevated zinc may be a sign of broader corrosion problems in a water supply system.

Appendix 7

Design of pipework systems in buildings

There are basically two types of water supply installation, the installation of a new system in a new building, or the installation of a new system in an existing building. In an existing building, where for example lead pipework may be found there may be an imperative to replace the existing pipes. Although it will be simplest to simply replace the existing pipes with new ones, this would not be wise as the circumstances of the building may have changed since the original system was installed and there may even be future plans for some form of development. If only a small section of pipe is to be replaced a straightforward substitution may be actable, but for a large scale replacement a careful consideration of the needs of the users of the building must be precede the new design.

Traditionally pipework design has been based upon fluid dynamics and traditional engineering assumptions. Although the physics of the systems has not changed over the years the expectations of users and the available appliances and fittings have changed. The increasing focus on demand management to address water consumption issues has led to the development of many water efficient appliances and terminal fittings which now need to be taken into consideration when designing a water supply system within a building. For example the typical flush volume of a WC has decreased from over 10 litres to less than 6 litres in most countries of the world over the last half century or so. On the topic of cisterns, in many countries where water supplies have been distributed to individual building's water supply cisterns there has been a trend towards direct mains connection. While this may provide higher outlet pressures and improve hygiene, it can increase the variability of supply and put an increased strain on the mains system operated by the water supplier. With high water use fittings and appliances, water in the supply pipes would be frequently used and there was a higher turn-over of water. In many buildings, energy efficiency was not a priority until recently, so pipes were often installed in unheated, draughty areas. However, in modern homes, especially those aiming to be zero-carbon consuming, the levels of insulation can lead to pipes being within a warm environment. Although insulation of pipes will help reduce thermal transfer between the surroundings and the water in the pipes, given long enough the water will attain the ambient temperature.

Plumbing Codes of Practice and some regulations may specify minimum supply pipe sizes. These are normally based upon traditional assumptions for the flow rates from fittings and discharge patterns, with the aim of not limiting the supply at any time. This can lead to considerable oversizing of systems and result in periods and areas of stagnation, low water turn-over and increased risk of Legionella. For cold water to be safe it needs to be maintained at a temperature below 20°C; although some countries may permit the temperature to rise to 25°C. In most buildings, this will require steps to be taken to reduce any heat transfer into the cold water system. One of the simplest methods is to always run hot pipes above cold pipes, so that convected heat rises away from the cold pipes. If possible hot and cold pipes should not be run in the same duct. All cold water pipes should be insulated to a level that is appropriate to the surroundings and ambient air temperatures. Once pipes are insulated any pipe markings will be obscured, so it is good practice to not only permanently identify the supply at the terminal fittings, but also at appropriate locations along the insulated pipes. This will help reduce the risk of any subsequent cross-connections and make subsequent supply identification easier.

For a new water supply system one of the first considerations will be material. Traditionally copper has been used in many installations, but as extraction, production and transport costs increase plumbers are often tempted to consider alternatives. Stainless steel has been used occasionally to replace copper, but its harder surface makes jointing a little more difficult and being metal it has similar pros and cons to copper. Since the introduction of plastic pipework a few decades ago, new types of plastic pipe are regularly being brought to market. Each type of plastic pipe has its own benefits and issues, so the selection has to take into account the availability of sizes and fittings, normal operating temperatures, peak operating temperatures and their durations, operating pressures related to the operating temperatures, UV stability, susceptibility to chlorination or other internal chemical treatments, oxygen permeability, external chemical resistance, colour, ease of jointing, flexibility, rigidity and support requirements. In addition, when using non-conductive pipework the issues of electrical earth bonding continuity may also need to be addressed. It is now common to have systems made of a mixture of materials. Flexible plastics are often used in 'threading' pipes through joists, ducts and walls and so on, but with any exposed pipework being completed in metal for appearance. More recently, composite pipes have become popular as they combine the flexibility of most plastics with the rigidity of metal and can be formed by-hand, and re-bent if needed. Composite pipes may need special fittings that seal on both the inside and outside of the pipe to produce a watertight seal. The internal seal may result in a reduction of pipe bore.

As many of these materials have different internal and external diameters, even for the same nominal size, interchanging systems and fittings is not recommended, so much care must be taken when carrying out remedial work to ensure that the correct fittings are used and that they are installed as the manufacturer intended. Pipework should never be used to support an appliance, fitting, cistern or tank; it must be installed so that it is not stressed and is free enough to accommodate the anticipated expansion and contraction without noise, but secure enough not to sag or rattle. Where fast closing valves are fitted, such as some solenoid valves, shock arrestors or other means to suppress water hammer may be needed.

The sizing of the pipework should be linked to the required volumes needed at the fittings or appliances. The aim should be that no water is stagnant in the system for more than a few minutes and that it is always delivered at a desirable and safe temperature. Traditionally a system would be one pipe that ran through the building with branches taken to individual outlets. Such a layout produces many sections of water that may not move for many days and may become heated by the surroundings. To eliminate these problems a different design is needed. Many of the principles that are used in 'clean-in-place' systems should be adopted to improve the hygiene of domestic and commercial plumbing systems. The principals include the total elimination of deadlegs, both at design stage and throughout the life of the system. It is quite common for terminal fittings to be removed and the associated pipework simply capped producing a section of pipe in which water will never flow. Such pipework should always be totally isolated or removed. The branches to terminal fittings or appliances can become deadlegs if the appliance or fitting is not frequently used. One way to eliminate this problem is to only use vestigial branch pipes with the connection to the outlet fitting being no more than three pipe diameters.

To ensure that water is regularly drawn through the pipework the layout of the pipes should ensure that the sentinel fitting or appliance at the end of the supply line is regularly used and discharges a significant volume of water. A frequently used WC cistern is normally ideal. Where it is not possible to designate a high use sentinel fitting forced or induced circulation should be considered. A pumped circulation system is a reliable proven solution, but it does consume energy. An alternative would be to use venturi effect tees that will induce a circulation in a water circuit if water is drawn off from one of these fittings.

Where pipework is designed to be used on an irregular basis, for example, for a seldom used machine or remote tap, consideration should also be given to the installation of an automatic flushing system that would discharge a sufficient volume of water on, say, a weekly basis to minimise the risk of long-term stagnation problems. Where the volumes discharged are significant, appropriate methods of reuse should be considered, for example, landscape irrigation, so that water is not simply sent to drain.

When the use of a water supply system is not in the manner for which it was designed, for example, due to changes in household occupation, staff levels or work patterns, a review of the system should be undertaken and changes made to ensure that the plumbing system can provide safe hot and cold water as currently required.

Additional references

British Standards Institute, Specifications for installations inside buildings conveying water for human consumption- Part 5: Operation and maintenance, BSI London 2012.
Clean in Place Best Practice Guidelines – Part III, Extra information on CIP For Smart Water Fund, August 2010. Hatlar Group Pty Ltd, **Level 6**, 22 William Street, Melbourne, Victoria 3000, Australia.

Appendix 8

Partial lead service line replacement with copper pipe and galvanic corrosion

Although a rigorous scientific analysis was never conducted, it has been traditionally assumed that partial replacement will eventually reduce lead in water versus leaving the whole lead service line intact. The common practice of only replacing the utility portion of lead service lines with copper pipe received newfound attention during implementation of the largest lead service line replacement program in US history in Washington DC. The $400 million project was initiated in 2004 as a step towards resolving the city's lead-in-water problems, but was eventually abandoned in 2008 partly due to data demonstrating extremely high spikes of lead in water for at least several months. Additionally, after analysing children's blood lead levels in the city, the US Centers for Disease Control and Prevention (CDC) concluded that partially replacing lead service lines did not decrease the risk of elevated blood lead levels ($\geq$10 µg/dL) associated with lead service line exposure, and even likely increased cases of elevated blood lead (Brown *et al.* 2011; Triantafyllidou & Edwards, 2011). The CDC study was the first to cast doubt over anticipated health benefits of the practice, although water utilities in the UK and the US have long warned of increased lead-in-water levels for an indeterminate length of time after partial replacements are conducted.

The increased lead in water after partial replacements can arise from a variety of mechanisms and can possibly be relatively short-term (days to weeks) or longer-term (months to years) in duration. Short-term problems are expected from disturbing the lead rust (i.e. corrosion scale) that has accumulated on the pipe over decades/centuries of use, and/or from creating metallic lead particles when the pipe is cut. Longer-term problems may arise due to galvanic (and/or deposition) corrosion of the remaining portion of lead pipe, when it becomes electrochemically connected to copper pipe or brass fittings. Based on the galvanic series, metallic lead typically serves as the anode of this galvanic cell and is therefore oxidized (i.e. corroded) to release Pb^{+2}, thereby contaminating the water. The copper pipe serves as the cathode, where the cathodic reaction (such as dissolved oxygen reduction) occurs over its surface. The production of Pb^{+2}, which is a Lewis acid, causes a local pH drop at the anode (i.e. the lead pipe surface or the point of metal to metal contact). In this situation lead leaching to water could increase due to higher corrosion rate and/or a lower pH at the surface of the lead material. Because low pH prevents formation of passive films on lead surfaces, this might contribute to self-perpetuation of the attack. This is especially true in cases where there is infrequent water flow in the pipe, as is common in lead service lines. Another potentially important, yet unappreciated, micro-galvanic phenomenon is deposition corrosion. Deposition corrosion can occur when cupric soluble ions released from the copper pipe are flowing through the remaining lead pipe, such that each site of copper deposition on the lead pipe may act as an individual galvanic cell and raise the water lead concentration (Triantafyllidou & Edwards, 2011).

In the UK, the lead: copper galvanic cell was thought to exacerbate the problem of plumbosolvency and give rise to increased/erratic levels of lead observed at the tap, even annulling any beneficial effects of reducing the length of lead pipe in the system (Triantafyllidou & Edwards, 2011). In the US, prior research on galvanic lead corrosion did not reach a consensus regarding the effect of partial lead service line replacement with copper pipe (Triantafyllidou & Edwards, 2011). Experiments of stagnant water conditions proved that under worst-case scenarios galvanic corrosion could significantly increase lead leaching for months or longer. Other experiments with pipe loops under continuous water flow suggested that galvanic corrosion was of short-duration, although quality assurance and quality control testing later revealed that release of particulate lead was potentially missed by the protocol. In 2011, a US EPA scientific advisory board examined the available literature, and noted that additional research to assess mechanisms and long-term impacts of partial lead service line replacements was a high priority. Subsequent experimental work (Giammar *et al.* 2011) confirmed that in some cases, galvanic corrosion of lead: copper pipe connections creates persistent and significant water lead contamination.

Considering the high financial burden on utilities to conduct partial lead service line replacements, anecdotal evidence of increased and erratic lead levels for an unspecified time thereafter and new documented health outcomes associated with this practice, full lead service line replacements should be encouraged in all cases. Alternatively, installing a non-metallic pipe to replace the utility portion of the lead service line or placing a dielectric between the remaining lead portion and the newly installed copper portion would eliminate direct galvanic corrosion concerns, although a dielectric would still allow deposition corrosion. Short-term problems from disturbing lead rust or cutting the lead pipe are unavoidable after any partial replacement, and homeowners should therefore be notified and advised to use water filters or flush their lines prior to water consumption for a period of time to avoid adverse health outcomes.

Because of lead accumulation in premise plumbing scales downstream of active lead sources, it confounds interpretation from a typical first-one-litre sampling scheme. Laminar flow effects make identifying localised slugs of lead from galvanic corrosion difficult to precisely define.

Additional references

Brown M. J., Raymond J., Homa D., Kennedy C. and Sinks T. (2011). Association between children's blood lead levels, lead service lines, and water disinfection, Washington, DC, 1998–2006. *Environ Res.*, **111**(1):67–74.

Triantafyllidou S. and Edwards M. (2011). Galvanic Corrosion after Simulated Small-Scale Partial Lead Service Line Replacements. *Journal AWWA*, **103**(9):85–99.

Giammar D., Wang Y., Jing H., Cantor A. and Welter G. (2011). Experimental Investigation of Lead Release During Connection of Lead and Copper Pipe. American Water Works Association Water Quality Technology Conference.

De Santis M. K, Welch M. M. and Schock M. R. (2009). Mineralogical evidence of galvanic corrosion in domestic drinking water. Proc. AWWA Water Quality Technology conference, Seattle, WA.

Appendix 9

Internal corrosion control in small supplies

Throughout the world there are a large number of very small water supplies that serve small communities or even individual dwellings and businesses. In Europe and the US, about 15% of the population utilise small or very small supplies. These supplies present a particular challenge to effective corrosion control for the following reasons:

- Raw water quality and quantity can be unreliable;
- Treatment is often minimal and inadequate;
- Technical awareness of those using or operating such supplies may be limited;
- Operation and maintenance of small systems is often lacking in rigour;
- Regulation and sampling may be limited or non-existent;
- Usage of unapproved or inappropriate plumbing materials is common.

The impact of poor corrosion control on such supplies and people using them is hard to quantify. Lead corrosion increases significantly in waters below pH 7, and for copper below pH 6.5 and an alkalinity of less than 60 mg/l $CaCO_3$. In Scotland, where raw waters are often very low in pH and alkalinity, compliance with the lower regulatory standard for pH of 6.5 is low, with over 25% of samples taken from the smallest (Type B) category of private water supplies not meeting this standard. Correspondingly, compliance with the standards for metals is also an issue – nearly 12% of samples failing the copper standard and 6.5% failing the lead standard.

Where waters are soft, often pH adjustment is not provided as part of the treatment process or is simply not maintained. For small supplies a contact vessel containing high alkalinity media such as limestone is often the simplest solution. These allow the remineralisation and pH adjustment of low alkalinity waters without the use of chemical dosing, which is harder to control. There are, however a number of considerations when using such contactors:

- The contact time must be set appropriately or the process will be ineffective;
- Some element of maintenance is still required as deposits (such as organic material and manganese) can build up on the media;
- Users and operators must be aware of the need to replenish media periodically.

Other options to reduce corrosion on small supplies include aeration, dosing an alkali to increase pH and dosing corrosion inhibitors such as polyphosphates or silicates. Where chemicals are dosed into a small water supply system, there are obvious risks if control of dosing is not adequate. On small systems this can be especially difficult where flows and water quality can change rapidly. Any pH adjustments should be made after chlorination.

With new supplies, metal leaching problems should be prevented by the use of modern materials and proper system design.

Additional reference

Drinking Water Quality Regulator for Scotland, Drinking Water Quality in Scotland – Annual Report 2010.

Part C

Check Lists and Criteria for Risk Assessment

Introduction

Part C comprises a series of Check Lists and Criteria that provide a working basis for the comprehensive risk assessment of corrosion control of a water supply system. None-the-less, the user should not hesitate to include any other factors considered to be relevant locally.

The criteria provide a basis for judging whether risks are "very high", "high", "moderate", "low" or "very low" by reference to available data. If no data is available, a "moderate" risk should be assumed by default.

The consequence of such risks will depend on both the nature of the risk (health related or aesthetic) and the population involved.

Not all the information, that is relevant to corrosion control, is risk assessed in a quantified manner, but will be relevant more broadly. This is reflected in the proformas that follow; some components have a risk assessment panel, others do not.

PROFORMA C1   ADMINISTRATIVE INFORMATION AND QUALITY ASSURANCE

1. Water utility contact details

Water utility	
Address	
Web-site	
Principal contact	
E-mail	
Direct dial	
Mobile	

2. Quality assurance

Item	Date	Signature
All data inputs are correct and up to date		
Liaison with all relevant internal personnel has been undertaken		
Liaison with all relevant external organisations has been undertaken		
Risk assessments completed with justifications		
Mitigation programme approved by Directors		
Reviews of mitigation programme scheduled after six months and after one year		
Next risk assessment scheduled		

PROFORMA C2   WATER SUPPLY SYSTEM SUMMARY

Water supply system	
Population served	
System description	*include distribution network materials: extent of cast iron, ductile iron, MDPE, asbestos cement, etc.*
Number of service connections	
Number of lead service lines	*to include customer-owned lead pipes*
Number of copper service lines	
Corrosion control history	*steps taken to minimise corrosion and when measures were taken; please specify pH conditions, the corrosion inhibitors used and their concentrations (including units of expression, e.g. mg/l P or PO$_4$)*
Plumbing Codes and Standards used in the system	

PROFORMA C3   WATER QUALITY: GENERAL RISK ASSESSMENT

Parameter: pH

Units: pH units	Date range of samples:
Sample type:	Number of samples:

Maximum:	Minimum:	Average:

Level of risk	Risk criteria	Risk based on maximum (tick)	Risk based on minimum (tick)	Risk based on average (tick)
Very high	<6.0			
High	<6.5			
Moderate	<7.0			
Low	7.1 to 7.5			
Very low	7.6 to 8.3			
Low*	8.4 to 9.5			
Moderate*	>9.5			

*in relation to dezincification, dependent on chloride/alkalinity ratio, otherwise "very low"

Consequence of risk: (consider general corrosivity)
Mitigation measures proposed:

Parameter: alkalinity

Units: mg/l($CaCO_3$)	Date range of samples:
Sample type:	Number of samples:

Maximum:	Minimum:	Average:

Mitigation measures proposed:*
*Although not subject to risk assessment in direct terms, action may be warranted to improve pH buffering characteristics of the water supplies

Parameter: chloride

Units: mg/l (Cl)	Date range of samples:
Sample type:	Number of samples:

Maximum:	Minimum:	Average:

Parameter: chloride/alkalinity ratio (dezincification)

Maximum:	Minimum:	Average:

Level of risk at $pH > 8.3$	Risk criteria	Risk based on maximum (tick)	Risk based on minimum (tick)	Risk based on average (tick)
High	>0.5			
Low	<0.5			

Consequence of risk: (consider potential for dezincification)
Mitigation measures proposed:

Parameter: sulphate

Units: mg/l(SO$_4$)	Date range of samples:
Sample type:	Number of samples:

Maximum:	Minimum:	Average:

Parameter: chloride/sulphate ratio (galvanic corrosion)

Maximum:	Minimum:	Average:

Level of risk	Risk criteria	Risk based on maximum (tick)	Risk based on minimum (tick)	Risk based on average (tick)
Moderate	>0.6			
Low	<0.6			

Consequence of risk: (consider potential for galvanic corrosion)
Mitigation measures proposed:

Parameter: total organic carbon

Units: mg/l(C)	Date range of samples:
Sample type:	Number of samples:

Maximum:	Minimum:	Average:

Level of risk	Risk criteria	Risk based on maximum (tick)	Risk based on minimum (tick)	Risk based on average (tick)
Very high	>5			
High	3 to 5			
Moderate	2 to 3			
Low	1 to 2			
Very low	<1			

Consequence of risk: (consider possibility of enhanced cuprosolvency and plumbosolvency)

Mitigation measures proposed:

Parameter: iron

Units:	µg/l(Fe)	Date range of samples:
Sample type:		Number of samples:

Maximum:	Minimum:	Average:

Level of risk	Risk criteria (µg/l)	Risk based on maximum (tick)	Risk based on minimum (tick)	Risk based on average (tick)
Very high	>300			
High	201 to 300			
Moderate	101 to 200			
Low	51 to 100			
Very low	<50			

Consequence of risk: (consider iron discolouration and potential for particulate lead problems)
Mitigation measures proposed:

Parameter: temperature

Units:	degrees Celsius	Date range of samples:	
Sample type:		Number of samples:	
Maximum:	Minimum:	Average:	

Over-all conclusions concerning water quality:

PROFORMA C4   PLUMBOSOLVENCY CONTROL

Regulatory compliance

Standard that applies:	Assessment method:
Sample type:	Frequency of assessment:

Survey results for the past five years

Year	Number of samples	Survey result

Any operational changes in the past five years that may have affected plumbosolvency control?

Operational monitoring

Standard that applies:	Assessment method:
Sample type:	Frequency of assessment:

Survey results for the past five years

Year	Number of samples	Survey result

Other monitoring investigations:

Lead service line replacement:

Number of lead service lines:	Number of full replacements:
Number of partial replacements:	Replacement material and jointing method:

Corrosion testing (specify type and results):

Analysis of corrosion deposits (specify method and results):

Modelling (specify type and results):

Use of corrosion inhibitors:

Type used and concentration	
Justification for the selection of the corrosion inhibitor	
Evidence that the use of the corrosion inhibitor is optimised	
Evidence that the pH conditions are optimised	

Over-all conclusions concerning the optimisation of plumbosolvency control:

PROFORMA C5   CUPROSOLVENCY CONTROL

Regulatory compliance

Standard that applies:	Assessment method:
Sample type:	Frequency of assessment:

Survey results for the past five years

Year	Number of samples	Survey result

Any operational changes in the past five years that may have affected cuprosolvency control?

Operational monitoring

Standard that applies:	Assessment method:
Sample type:	Frequency of assessment:

Survey results for the past five years

Year	Number of samples	Survey result

Consumer complaints:

Evidence of any pitting corrosion:

Corrosion testing (specify type and results):

Analysis of corrosion deposits (specify method and results):

Modelling (specify type and results):

Use of corrosion inhibitors:

Type used and concentration	
Justification for the selection of the corrosion inhibitor	
Evidence that the use of the corrosion inhibitor is optimised	
Evidence that the pH conditions are optimised	

Over-all conclusions concerning the optimisation of cuprosolvency control:

PROFORMA C6   IRON CORROSION CONTROL

Regulatory compliance

Standard that applies:	Assessment method:
Sample type:	Frequency of assessment:

Survey results for the past five years

Year	Number of samples	Survey result

Any operational changes in the past five years that may have affected iron corrosion control?

Operational monitoring

Standard that applies:	Assessment method:
Sample type:	Frequency of assessment:

Survey results for the past five years

Year	Number of samples	Survey result

Other monitoring investigations:

Mains rehabilitation:

Total length of water mains:	Length of mains rehabilitated:
Planned length of mains to be rehabilitated:	Planned completion date:

Corrosion testing (specify type and results):

Analysis of corrosion deposits (specify method and results):

Network modelling (specify type and results):

Use of corrosion inhibitors:

Type used and concentration	
Justification for the selection of the corrosion inhibitor	
Evidence that the use of the corrosion inhibitor is optimised	
Evidence that the pH conditions are optimised	

Over-all conclusions concerning the optimisation of iron corrosion control:

PROFORMA C7   IMPROVEMENT PROGRAMME

No	Issue to be addressed	Improvement measures	Completion date	Person responsible
1				
2				
3				
4				
5				
6				
7				
8				
9				
10				

Authorised by:	Signed:	Date:

9 781780 404547